T0202721

Bionic Optimization in Structural Design

Rolf Steinbuch • Simon Gekeler

Editors

Bionic Optimization in Structural Design

Stochastically Based Methods to Improve the Performance of Parts and Assemblies

 Springer

Editors
Rolf Steinbuch
Reutlingen University
Reutlingen
Germany

Simon Gekeler
Reutlingen University
Reutlingen
Germany

ISBN 978-3-662-51605-8   ISBN 978-3-662-46596-7   (eBook)
DOI 10.1007/978-3-662-46596-7

Springer Heidelberg New York Dordrecht London
© Springer-Verlag Berlin Heidelberg 2016
Softcover re-print of the Hardcover 1st edition 2016

Printed on acid-free paper

Springer-Verlag GmbH Berlin Heidelberg is part of Springer Science+Business Media (www.springer.com)

Preface

Bionics has become more and more popular during the last few decades. Many engineering problems are now solved by copying solutions found in nature. Especially the broad field of optimization has been inspired by the variety of methods to accomplish tasks that can be observed in nature. Popularly known examples include the strategies that ant colonies use to reduce their transport distances to feed their always hungry population, the dynamics of swarms of birds or fishes, and even replication of the brain's learning and adapting to different challenges.

Over more than a decade, we have been studying Bionic Optimization at the Reutlingen Research Institute (RRI). After early attempts to design optimization solutions using parameterized CAD-systems and evolutionary strategies, our field of interest became broader. Our work taught us how the different bionic optimization strategies might be applied, which strong points and which weaknesses they exhibited, and where they might be powerful and where inappropriate.

During a series of joint research projects with different partners and supported by the German government and other sponsors, we studied many aspects of these techniques. Additionally, the interest of the scientific community in Bionic Optimization is increasing along with the fuller understanding of how engineering can be influenced by non-deterministic phenomena. In this book we intend to give an introduction to the use of Bionic Optimization in structural design. Readers should be enabled to begin applying these nature inspired procedures. Furthermore, hints about the implementation, useful parameter combinations, and criteria to accelerate the processes are included.

To formulate most bionic optimization processes, scientists have attempted to base the strategies on a strong and reproducible theoretical foundation. On the other hand, most of these methods are so easy to understand that we realize they are working even if we decline to base them on a strict mathematical background. In this book we decided to explain the basic principles, show examples that are easy to understand, and list easily reproducible pseudocode to help new users to start working immediately. Comments on meaningful parameter combinations and warnings on problems and critical configurations may motivate readers to verify whether our proposals are justified, or if they can be expanded to broader regimes.

v

The work presented in this book mostly is a re-composition of different papers, theses, work reports, and presentations written throughout the last decade. The authors are former or current students at Reutlingen University, colleagues at the RRI, people who like working in Bionics, and young engineers who had, and have, plenty of ideas and are not too easily frustrated by flops. We have been following many tangents, have done thousands of studies, and have found solutions to many questions, but sometimes have failed to find the answers to others.

We begin with basic definitions and motivations, giving simple examples, and explaining how to set up an optimization environment. Some more elaborate applications then exhibit the power of these methods. Finally, a discussion about the future developments indicates how we expect optimization to be used in the future.

All this work would not have been possible without the support of many different sponsors. Besides the financial support of the German government in some research projects, many software companies and manufacturing enterprises gave us the opportunity to scan the wide range of bionic optimization in industry. We recognize their help, the fruitful discussions, and the generous handling of the licensing of the software packages. Additionally, we would like to express our gratitude to the heads of Reutlingen University, the RRI, and the faculty of engineering all of whom gave us access to space, time, and nearly endless computing power. We want to express our gratitude to Springer, especially Mrs. Eva Hestermann-Beyerle and her staff, who have helped so much to transform the collection of many different papers in different formats into one readable book.

Reutlingen, Germany   Simon Gekeler
April 2015   Rolf Steinbuch

Contents

About the Editors and Authors

About the Editors

Simon Gekeler, from Reutlingen, Germany, studied Mechanical Engineering at Reutlingen University and finished his Master thesis in 2012. While preparing his Master thesis and during the following years as a research assistant at Reutlingen Research Institute (RRI), he did research in methods of structural optimization, sensitivity analysis and evaluation of design robustness and design reliability.

Rolf Steinbuch, from Stuttgart, Germany, studied Mathematics and Physics at the University of Ulm. After 5 years with Siemens Power Stations, he moved to Mercedes-Benz. There he helped to introduce non-linear simulation into design processes. Since 1993 he is responsible for Numerical Structural Mechanics at Reutlingen University. His research focusses on optimization, acoustics and nonlinear problems.

Contributing Authors

Stephan Brieger, from Nürtingen, Germany, studied Mechanical Engineering at Reutlingen University and finished his Master thesis in 2006. While preparing his Master thesis and during the following 2 years as a research assistant at Reutlingen Research Institute (RRI), he did research in methods of structural optimization. Since 2009, he has been employed at Kolt Engineering in Böblingen as a Project Engineer, responsible for dynamics and vibration analysis.

Dmitrii Burovikhin, from Kotlas, Russia, completed his Bachelor and Master degrees in Mechanical Engineering from 2006 to 2012 at St. Petersburg State Polytechnic University. He finished a second Master degree in Mechanical Engineering at Reutlingen University in 2015. Within the RRI simulation team, he is responsible for the design of optimization tools driving CAE-systems.

Nico Esslinger, from Oberndorf, Germany, finished his Master of Mechanical Engineering at Reutlingen University in 2015. Since 2014, he has been a research assistant at Reutlingen

University with the Reutlingen Research Institute. Currently, he is working in the field of Multi-Objective Optimization.

Andreas Fasold-Schmid, from Reutlingen, Germany, completed his Bachelor in Mechanical Engineering at Reutlingen University in 2013. He is now finishing his Masterthesis there in the field of optimization.

Oskar Glück, from Reutlingen, Germany, studied Mechanical Engineering at Reutlingen University. Currently he is in the Masterclass for Mechanical Engineering at Reutlingen University and is part of the RRI Simulation group, responsible for process acceleration.

Iryna Kmitina, from Dnipropetrowsk, Ukraine, studied Control and Automation at the National Mining University of Ukraine from 2000 to 2005. She has worked as a research assistant at the National Mining University of Ukraine, Department of Automation and Computer Systems. Since 2013 she has been working as a research assistant at Reutlingen University with the Reutlingen Research Institute in the field of metal forming process optimization.

Julian Pandtle, from Reutlingen, Germany, studied Mechanical Engineering at Reutlingen University and finished his Master thesis in 2011. During his thesis he worked with the RRI dealing with EVO and PSO methodologies. Since 2011 he has been working at WAFIOS AG in Reutlingen, responsible for nonlinear dynamics and metal forming.

Tatiana Popova, from St. Petersburg, Russia, completed her Master of Mechanical Engineering in St. Petersburg, before joining the RRI. At Reutlingen University, she completed her second Masters and has been working with metal forming process optimization.

Frank Schweickert, from Kirchheim unter Teck, Germany, studied Mechanical Engineering at Reutlingen University, and finished his Bachelor thesis in 2014. Currently he is pursuing his Masters degree in Biomimetics: Mobile Systems, at the University of Applied Sciences Bremen.

Ashish Srivastava, from Lucknow, India, received his Bachelor degree in Mechanical Engineering at Uttar Pradesh Technical University in 2010, then continued at the University of Duisburg-Essen (Germany) to complete his Master of Science in Computational Mechanics in 2014. He has worked at the University Duisburg-Essen and the Fraunhofer-Gesellschaft (SCAI) and is now at RRI in the field of Robust Design Optimization.

Christoph Widmann, from Reutlingen, Germany, studied Mechanical Engineering at Reutlingen University and finished his Masters thesis in 2012. During his thesis, he worked with the RRI specializing in Neural Nets. Since 2012, he has been working at WAFIOS AG in Reutlingen, where he is responsible for nonlinear dynamics, metal forming and vibration analysis.

Chapter 1
Motivation

Rolf Steinbuch

Since human beings started to work consciously with their environment, they have tried to improve the world they were living in. Early use of tools, increasing quality of these tools, use of new materials, fabrication of clay pots, and heat treatment of metals: all these were early steps of optimization. But even on lower levels of life than human beings or human society, we find optimization processes. The organization of a herd of buffalos to face their enemies, the coordinated strategies of these enemies to isolate some of the herd's members, and the organization of bird swarms on their long flights to their winter quarters: all these social interactions are optimized strategies of long learning processes, most of them the result of a kind of collective intelligence acquired during long selection periods.

1.1 A Short Historical Look at Optimization

In consequence it is not surprising to find optimization approaches in more highly organized human societies, focusing, for example, not only on the organization of social life but also on craftsmanship as well. Qualified professionals learn, try, fail, and improve until they are capable of performing their craft to certain perfection. And then new workers come, with the desire to surpass their antecessors, and create even better ideas and products. With increased productivity and the shorter lifetime cycles of industrial production, the need to deliver higher qualities in shorter times has become a continuous challenge. Today optimization is an inherent part of the industrial process. Since engineering, especially the design of machinery, started to

R. Steinbuch (✉)
Hochschule Reutlingen, Reutlingen Research Institute, Alteburgstraße 150, 72762 Reutlingen, Germany
e-mail: Rolf.Steinbuch@Reutlingen-University.DE

© Springer-Verlag Berlin Heidelberg 2016
R. Steinbuch, S. Gekeler (eds.), *Bionic Optimization in Structural Design*,
DOI 10.1007/978-3-662-46596-7_1

1

become a discipline, more than merely an appendix of the manufacturing process, the task of optimization has been incorporated within its precincts.

1.1.1   Optimization in Engineering History

The founding days of Technical Mechanics, starting with the analysis of simple rods and beams, enabled engineers to predict the load carrying capability of a theoretical part and to select acceptable variants. At these early stages, an essential part of mechanical and civil engineering was devoted to finding methods, formulas, and predictions of the response of systems and structures. Engineers used these formulas to discover better solutions. Optimization might be regarded at least as one of the central items in mechanical engineering. Good engineers understand the processes they deal with, improve them, apply the relevant theoretical approaches, work out the essential consequences of the theory, and interpret them in an appropriate way. Following this approach, which is based on abstract thinking, the optimization is then transferred to the physical models. Through this process, engineers analyzed why the models did not work as expected, improved their understanding of the processes, and then designed new and better models. Parallel to this development, the efficiency of mathematical methods became more and more important. Among the central difficulties at that time was dealing with non-trivial formula, solving problems with more than two or three unknowns, studies of processes in time and space, and many other mathematical problems that required powerful handling of numerical tasks.

1.1.2   Finding Relevant Numbers in Engineering

Early on, finding the correct numbers for specific problems became a central challenge in the mathematical analysis of engineering problems, so there were many attempts to build calculators. Charles Babbage's difference engine and analytical engine, built at the beginning of the nineteenth century, was among the first and certainly among the most famous. But it was not until the 1930s that various developers, using electric current instead of mechanical contacts as leading technology, succeeded to produce relatively fast and reliable computers. The development of the transistor in the late 1940s allowed for the assembly of computers which were not built with relays or electronic valves and which were both very fast and very reliable compared to their predecessors. Up to today, we do not see any limits to the growing calculation capacity of these transistorized computers. In consequence, we are able to solve large problems with many unknowns in a short time, and this has caused Technical Mechanics to lose much of its frustrating aspects to engineers.

1.1.3 High Level Mechanical Methods

Parallel to the development of computers, new methods in mechanics arose. Beginning with early steps in the nineteenth century, Walter Ritz (1908) and Boris Galerkin (1915) proposed a method to solve structural problems that might be essentially more complex than the ones handled by the classical formula (Fig. 1.1). Richard Courant (c. 1923) was the first mathematician to understand the potential of their proposal, but development of these ideas was limited by lack of computing power. However, during the Second World War and in the years following it, scientists started to propose variants of these original ideas, which we know today as the Finite Element Method (FEM). Parallel to the FEM, the Boundary Element Method (BEM), often associated with Erich Trefftz, was developed and became an important tool in many engineering applications.

In the first years of use, the industrial application of both FEM and BEM was restricted to applications in which minimizing cost was a lower priority. So, air- and spacecraft, military weapons, nuclear industries, and some high level vehicle applications were using the then expensive numerical tools. In addition to the expense of computing power, up to the 1980s, the large effort to define and to enter the geometrical properties such as nodes and elements in FEM reduced the applications to simple problems and isolated studies. Consequently, in the early 1990s the meshing of a motor-head took about 6 months, not taking into account the other 6 months required to develop the wireframe CAD-model that served as basis for this FEM-model.

1.1.4 Drop of Hardware Costs and Better CAD Systems

In the 1990s two essential developments took place. 3D-CAD-Systems using solid models were developed. They required many fast and well performing local graphical systems be installed on powerful workstations. Since the workstations

Fig. 1.1 Walter Ritz (Ritz), Boris Galerkin (Galerkin), Richard Courant (Courant), Erich Trefftz (Trefftz)

a)   b)

c)   d)

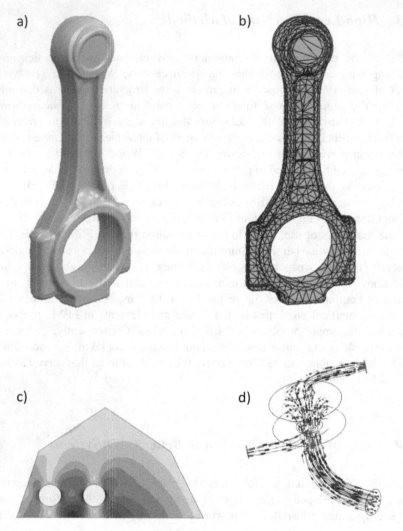

Fig. 1.2 Examples of simulation tools. (**a**) CAD-model. (**b**) FEM-mesh. (**c**) Stresses around tunnels in a mountain (BEM). (**d**) Flow through a nozzle pair (FVM)

of individual engineers could perform much of the computation, there was no longer a need for large central mainframes. More importantly, the FEM-meshes could be easily derived from the 3D-CAD-Models. Only with these advancements would FEM became a tool available to more diverse segments of industry as well.

The new, less expensive application of FEM and other simulation systems, such as BEM or the Finite Volume Method (FVM) for fluid mechanics problems, opened up possibilities to apply field-integrated optimization (Fig. 1.2). It was once acceptable to spend hours building a model. But it is far too expensive to spend many hours on the repeated process than to build and study variants of the initial design.

Table 1.1 Development of PC hardware performance and prices in 18 years

Year	RAM (Mbyte)	Disc (Mbyte)	CPU (MHz)	Price (€)
1994	8	400	80	2000
2012	24,000	4,000,000	12×3200	2000
Growth	3000	10,000	480	1
Growth/year	2.72	3.16	2.16	1

RAM: core memory; Disc: storage on any type of disc; CPU: processor speed, # of operations/second (frequency) multiplied by # of cores

So in the late 1990, the first CAD- and FEM systems were introduced that did essential parts of the optimization cycle automatically.

Because the increased power of PC's reduced the price of computation even more, application of optimization methods was no longer restricted by the hardware costs. Simultaneously, the power of such very common systems grew exponentially, as we see in Table 1.1. The frequently mentioned Moore's law seems to still be valid. The performance of computers is more than doubling every year while the prices do not grow or even drop.

1.2 Optimization and Simulation as Part of the Virtual Product Development

The expanding use of simulation tools in the design of components has caused the transformation of development from mostly hands-on experience and experiment-based to one with a new nature, namely, Virtual Product Development. Using simulation tools, especially FEM for structural problems, Multi-Body-Systems (MBS) for kinematic studies, and Finite Volumes (FVM) in Computational Fluid Dynamic (CFD) for fluid mechanics problems, have enabled the analysis of many variants, the isolation of critical details, and the improvement of basic proposals until solutions had been found that perform essentially better than their predecessors. The costs of even large numbers of studies have been decreasing to such small amounts that there is no reason not to check yet more variants, try uncommon ideas, and search for designs in regions of the parameter space that would never have been checked in an experimental configuration, as the costs would be prohibitive.

These studies covering expanded segments of the free parameter space of the component's design cried out for automation. Why should an engineer repeatedly produce variants when the CAE-system could be doing so by itself?

1.3   Optimization in Nature

Darwin's discovery of the evolution and adaption of living structures to their
environment was one of the most shocking events in the history of human science
and culture. The fact that we are not an extraordinary species standing high above
the rest of our cohabitants, but that we are only one of many modifications of one
basic design which had been developing for more than 3 billions of years caused
widespread and heated debates, even leading to the prediction that Darwin's
followers would be consigned to hell. On the other hand, people more open to
scientific understanding soon learned that a powerful tool was inherent within the
principles of species crossing, mutation, selection, and adaption. Some of them
were thus lead to observe nature's design process and to attempt to adapt and copy it
to their specific tasks. They learned that some feathers at the end of a bird's wing
improved the stability at higher flight velocities (Fig. 1.3a), different types of eyes
made different species fit to recognize prey or predator, and smooth connections
between trees trunks and branches reduced the danger of breaking (Fig. 1.3b). Many
other examples are known, such as in the organization of social systems from
human communities to ant colonies which perform efficient travels to keep their
large population fed and housed. Looking at nature from a designer's point of view,
we see nothing but the preliminary results of never-ending optimization processes.

We understand today that, from a certain point of view, all living beings are the
results of optimization processes. Common to all these processes is that they have
been ongoing for a very long time, using many cycles and producing many less
successful variants. However, the different species would not have been able to
survive, if they had not been able to adapt to the ever-changing environment. In
fact, fascinating animals such as the famous dinosaurs, may have died out due to
their lacking the capability to adapt to the massive changes in their world. One
central question of the bionic evolution is still under debate by the scientific
community: are we observing the survival of the fittest or of the least unfit? As

a)   b)

Fig. 1.3 Examples of optimization in engineering and nature. (**a**) Winglet of an airplane inspired
by birds of prey. (**b**) Connection of branches to a trunk

this is a rather abstract position related to our topic of Bionic Optimization, we do not want to deepen this debate.

1.4   Terms and Definitions in Optimization

To discuss optimization effectively in the next chapters, we have to agree on a common language, i.e. the use of the same terms for the same phenomena. As optimization research is done by various groups within various and diverse scientific fields, and also in different regions of the earth, there is the real danger to get confused as the meanings of terms may diverge from group to group. Thus, we must clarify the set of terms used in this book. Most people involved in optimization accept that for an optimization study:

- We need a given *goal* or *objective z*.
- This objective z depends on a set of *free parameters* $p_1, p_2, \ldots p_n$.
- *Limits* and *constraints* are given for the parameters values.
- There are *restrictions* of the parameter combinations to avoid unacceptable solutions.
- We seek to find the *maximum* (or *minimum*) of $z(p_1, p_2, \ldots p_n)$.

To better define our terminology, we use the following conventions and findings:

- The *objective* or *goal* must be defined *à priori* and uniquely. Changing the definition of the goal is not allowed, as this poses a new question and requires a new optimization process.
- We need to define all free parameters and their acceptable *value ranges* we might modify during the optimization studies.
- This *value ranges* or *parameter range* is the span of the free parameters given by lower and upper limits. Generally it should be a continuous interval or a range of integer numbers.
- The fewer free parameters we must take into account, the faster the optimization advances. Consequently, accepting some parameters as fixed reduces the solution space and accelerates the process.
- Restrictions, such as unacceptable system responses or infeasible geometry, must be taken into account. But sometimes restrictions limit the ranges of parameters to be searched. Such barriers have the potential to prevent the optimization process from entering interesting regions.
- Finding the maximum of $z(p_1, p_2, \ldots p_n)$ is the same process as finding the minimum of the negative goal $-z(p_1, p_2, \ldots p_n)$. There is no need to distinguish between the search of maxima or minima.

Gradient based optimization methods are the most popular ways to find improvements of given situations. From an initial position, the derivatives of the objective

$z(p_1, p_2, \ldots p_n)$ with respect to the free parameters are determined. As the derivatives are often not to be found by analytical ways they are approximated by

$$\frac{\partial z}{\partial p_k} \approx \frac{z(p_1, p_2, \ldots p_k + \Delta p, \ldots p_n) - z(p_1, p_2, \ldots p_k - \Delta p, \ldots p_n)}{2 \Delta p} \quad (1.1)$$

The column of these derivatives defines the gradient:

$$\nabla z = \left(\frac{\partial z}{\partial p_1}, \frac{\partial z}{\partial p_2}, \ldots \frac{\partial z}{\partial p_n} \right)^T \quad (1.2)$$

Jumping along this gradient, for example, by using a line search method such as Sequential Quadratic Programming (SQP) (Bonnans et al. 2006), or any related method, has the tendency to find the next local maximum in a small number of steps, as long as the search starts not too far away from this local maximum (Fig. 1.4).

Optimization using this climbing of the ascent of the gradient is often labelled as the Gradient Method or included in the set of deterministic optimization methods. Here each step is determined by the selection of the starting point. Unfortunately, the determination of the gradient requires $2n + 1$ function evaluations per iteration, which may be an extended effort if the number of parameters is large and the hill not shaped nicely.

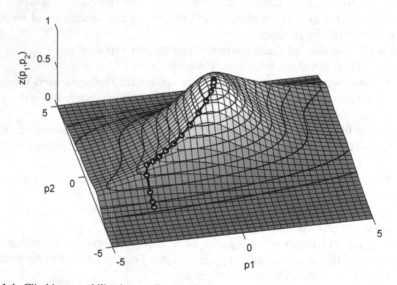

Fig. 1.4 Climbing up a hill using gradient methods

1.5   Why Bionic Optimization?

Deterministic methods, such as gradient climbing discussed in Sect. 1.4, fail as soon as there are many local hilltops to climb. Only the next local maximum is found if problems occur such as the one shown in Fig. 1.5.

An alternative is using purely stochastic searches, which may consist of randomly placed points into the parameter space. They guarantee discovery of the optimum, but only if we allow for very large numbers of trials. For real engineering applications, they are far too slow. A more powerful class of methods produces some random or motivated initial points into the parameter space and uses them as starting points for a gradient search. As long as the problem is of limited difficulty and does not have too many local optima, this might be a successful strategy. For problems that are more difficult to handle, the bionic methods presented in Chap. 2 prove to be more successful. They combine randomness and qualified search and have a sufficient potential to cover large regions of high-dimensional parameter spaces. Some randomly or intentionally placed initial designs are used to start an exploration of the parameter space and propose designs that might be outstanding, if not even the best. We discuss some of the Bionic Optimization methods in Chap. 2 and give the basic ideas, examples of applications, and sketches of program structures.

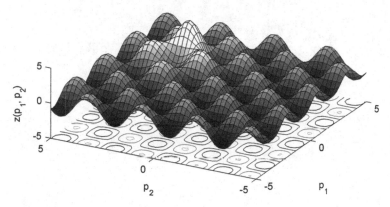

Fig. 1.5 Multi-Hill landscape with many local optima

References

Bonnans, J.-F., Gilbert, J. C., Lemarechal, C., Sagastizábal, C. A. (2006). *Numerical optimization – theoretical and practical aspects*. s.l. Berlin: Springer.

Courant. *Picture from: Courant, Reid. C, 1996*. s.l. Springer.

Galerkin, B. G. (1915). On electrical circuits for the approximate solution of the Laplace equation. *Vestnik Inzhenerov i Tekhnikov, 19*, 897–908.

Galerkin. *Boris Galerkin, picture from: Wikimedia Commons (27.04.15)*. http://www-groups.dcs. st-and.ac.uk/~history/Mathematicians/Galerkin.html

Ritz, W. (1908). Über eine neue Methode zur Lösung gewisser Variationsprobleme der mathematischen Physik. *Journal für die reine und angewandte Mathematik. 1908, 135*, 1–61.

Ritz. *Walter Ritz, picture from: Wikimedia Commons (27.04.15)*. http://www.archive.org/details/ gesammeltewerkew003778mbp

Trefftz. *Erich Trefftz, picture from: Wikimedia Commons (27.04.15)*. Original: Journal "Deutsche Mathematik", Vol. 2, Issue 5 (22 Dec 1937), p. 582; included in an obituary "Erich Trefftz" by Fritz Rehbock (p. 581–586).

Chapter 2
Bionic Optimization Strategies

Rolf Steinbuch, Julian Pandtle, Simon Gekeler, Tatiana Popova,
Frank Schweickert, Christoph Widmann, and Stephan Brieger

The realization that nature is a continuous optimization process founded a new engineering discipline. Beginning in the 1960s, various schools began to study and to reproduce bionic processes to improve previous optimization solutions. One of the most well-known examples is the winglet, an extension to the wings of aircrafts to stabilize the flow around the end of the wing. Another application is the sandwich structures of tailored blanks, where a sheet of material is subdivided into different layers. Only the outer ones, which are most important to the stiffness, are made of heavy and expensive metals. The filler, which only needs to keep the metal sheets separated by a certain distance, is made from less expensive and light-weight material. These blanks, especially honeycombs, are extremely stiff structures composed of minimal amounts of material. Unfortunately, their properties are very non-isotropic, so their use as load-carrying materials must be done with care and understanding.

Different organizations specialize in bionics, such as the International Society of Bionic Engineering (ISBE, http://www.isbe-online.org/, retrieved 15.04.2015) or

R. Steinbuch (✉) • S. Gekeler • T. Popova
Hochschule Reutlingen, Reutlingen Research Institute, Alteburgstraße 150, 72762 Reutlingen, Germany
e-mail: Rolf.Steinbuch@Reutlingen-University.DE; Simon.Gekeler@Reutlingen-University.DE; Tatiana.Popova@Reutlingen-University.DE

J. Pandtle • C. Widmann
WAFIOS AG, Silberburgstraße 5, 72764 Reutlingen, Germany
e-mail: J.Pandtle@wafios.de; C.Widmann@wafios.de

F. Schweickert
Hochschule Bremen City University of Applied Sciences, Neustadtswall 30, 28199 Bremen, Germany
e-mail: fschweickert@stud.hs-bremen.de

S. Brieger
KOLT Engineering GmbH, Schickardstraße 32, 71034 Böblingen, Germany
e-mail: S.Brieger@kolt.de

© Springer-Verlag Berlin Heidelberg 2016
R. Steinbuch, S. Gekeler (eds.), *Bionic Optimization in Structural Design*,
DOI 10.1007/978-3-662-46596-7_2

11

national networks such as BIOKON (http://www.biokon.de, retrieved 15.04.2015) in Germany. At Georgia Tech University (http://www.ece.gatech.edu/research/labs/gt-bionics, retrieved 15.04.2015) is a lab devoted to bionics; many other universities are working with bionic questions around the world. So today, learning optimization from nature is a broadly accepted method.

In this chapter we introduce methods to improve mechanical designs by bionic methods. In most cases we assume that a general idea of the part or system is given by a set of data or parameters. Our task is to modify these free parameters so that a given goal or objective is optimized without violation of any of the existing restrictions.

2.1 Evolutionary Optimization

Julian Pandtle

Evolutionary Optimization copies the way bionic beings reproduce and adapt to a changing environment. The combination of the properties of different individuals modifies the skills of their kids. In every generation the best kids will survive better and create another generation of kids, which leads to a higher fitness of the overall population (survival of the fittest). Just as in biological evolution, the properties that are transferred to the generations' kids derive from random mutation. The parents' original properties are changed, dependent upon the mutation radius, without knowing where the mutation leads. As the children within a reproducing population are never identical, some of them will be better suited to adapt to the environmental challenges. Over the sequence of generations, their genetic code may become dominant within the population.

2.1.1 Terms and Definitions

Here the most important terms in Evolutionary Optimization are listed. There are many other terms used in various papers, as there is no generally accepted vocabulary. Users are advised to check carefully the definitions when reading papers from various authors.

Individuals are the specific elements within the sets of parents and kids.

Generation is one step in the evolutionary process. It is given by a set of parents or individuals. The creation of a new set of kids or new individuals defines the genesis of a new generation.

Mutation is the modification of the parameter values of an individual. Mutation may happen in many ways (Rechenberg 1994; Gen and Cheng 2000; Steinbuch 2010). Every parameter may be changed by a random value. Some parameters may be changed in some correlated way. In some cases only a part of the parameters is changed in every generation. There are infinite possibilities for mutations, so the preferred types of mutation have to be checked carefully, until sufficient experience determines otherwise. In this book we mostly mutate all free parameters by the same mutation principle. We add the given mutation radius of each parameter weighted by a random number $-1 \leq r \, and \leq 1$ to the initial value of the parameter.

The *mutation radius r* defines the amount that a parameter may be changed in a mutation step. This value can be the same for all parameters or different for some or all parameters. Often it is not given as an absolute value but as the percentage of the total allowed range of a parameter. So the generation of the kids can generally be defined by different approaches:

- Random mutation:
  The *n* kids are randomly distributed within a restricting mutation radius around their initial set of parameters (Fig. 2.1a).
- Mutation based on probability distribution:
  The *n* kids are placed according to a distribution function around their initial parameter set. Often we use a normal distribution (Fig. 2.1b), where the mutation radius defines the standard deviation.

There are many other possibilities to do this mutation phase. For most of the approaches, it is possible to change, evaluate and optimize the mutation radius over the sequence of generations. It should be mentioned that the use of a small mutation radius reduces the process to a local search for an optimum. On the other hand, a too large mutation radius yields a similar performance as a pure random search.

The *number of parents* should be sufficiently large to cover some or many possible parameter combinations. Its value should generally not fall short of 0.5–5 times the number of free parameters.

The *number of kids* covers the parameter space. Again, a large number is preferred. Common experience, which may not be applicable in every case, proposes the number of kids to be 2–5 times the number of parents. See Sect. 3.3.1 for a discussion of the selection of well performing combinations of these input data.

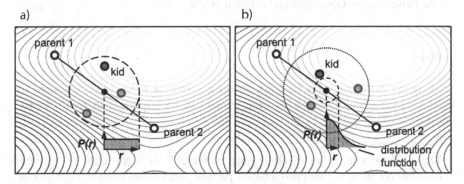

Fig. 2.1 Different mutation methods in Evolutionary Optimization. (**a**) Random mutation radius. (**b**) Mutation based on normal distribution

Pairing is the selection of two individuals of the parent generation to produce one common child.

Crossing, the way by which two parents define the properties of one common child, may happen in different ways. One of the first ideas is to average the parameter values of both parents (Fig. 2.1). Another type of crossing could be taking randomly one parameter from one parent only. Some weighted average of the two parents' data, e.g. preferring the better parent, is also a possibility (Gen and Cheng 2000; Steinbuch 2010).

The quality of each individual is prescribed by its *fitness*. It can depend on one single value or be the weighted result of different optimization interests (cf. Sect. 6.2).

Selection determines which kids of the current generation (including their parents or not) should be the parents of the next generation. Often selection is done by only taking the best kids as new parents. Experience suggests that at least some lesser performing kids should be parents as well. For the sake of simplicity and comparability, here we allow the old parents to be parents in following generations and take the best of the total of old parents and kids to be parents in the next generation. Sometimes it proves efficient to define a restricted lifetime for such parents, given by a maximum number of generations.

2.1.2 Description of the Evolutionary Strategy

In the first step a number of individuals are randomly distributed within the design space. We determine their fitness, and they are the parents in the first generation.

For proliferation, the selected individuals are arranged in pairs. Every pair generates one kid only by combining the properties of their parameter sets. This crossed parameter set is subject to some random modification prescribed by the *mutation*. Then best kids and parents are chosen to be parents of the next generation. The poorer performing kids are removed from the gene pool. A pseudocode for a simple Evolutionary Optimization is shown in Table 2.1.

Variants

As there are different ways to cross, select and mutate individuals, we want to discuss selected modifications that are thought to be important in Evolutionary Optimization.

– Pairing Variants:
  There are several ways of combining individuals to form a set of parents. Possibilities include randomly chosen pairs of individuals from the number of prospective parents, or the pairwise combination of individuals sorted by their fitness (combining the fittest pair, the second fittest pair, see Fig. 2.2).

Table 2.1   Pseudocode: Evolutionary Optimization

Initial: Define parameters:
  – Number of parents, kids, generations
  – Shall parents survive?
  – Crossing scheme, mutation radius
  – Define *nparents* parents; evaluate their performance and restriction violation

Start *ngenerations* loop:
  – Define *nkids* pairs of parents
  – Cross their properties
  – Mutate the properties to define kids
  – Evaluate the kids' performance and restriction violation
  – Select the next generation's parents from the kids (including parents?)

End loop

Stop

Fig. 2.2 Schematic diagram of Evolutionary Optimization: six initial individuals, four parents, six kids, over three generations

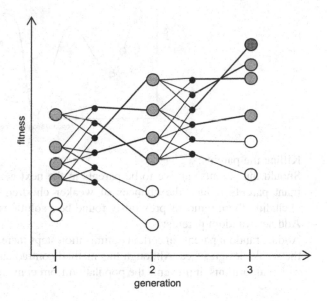

A possibility to affect the chosen sets of partners is combination based on the similarity of their properties, just as for many organisms in nature. The choice of partners depends strongly on similar interests, habitat, level of intelligence, attractiveness and so on. This kind of pairing can lead to a faster convergence because it concentrates the search only near optimal regions, and neglects the spaces in-between (Fig. 2.3a). However, any optima in the neglected spaces will not be found (Fig. 2.3b). Nevertheless, this method offers an interesting possibility to force the optimization to run in a specific direction. Users can include basic knowledge of the problem in the start of the optimization to affect the kinds of optima they find.

Fig. 2.3 Examples of
pairing individuals with
limited distance. (**a**) Pairing
with limited distance lead to
fast convergence.
(**b**) Pairing with limited
distance failed to find
optimum

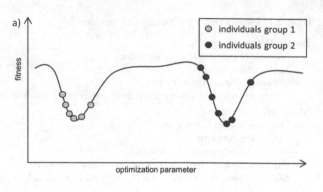

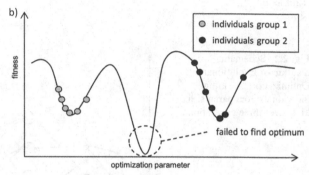

- Killing the parents:
  Should the parents survive to be parents in the next generation as well? Dominant parents reduce the chances of weaker children to find their way, but excluding them removes previously found best solutions.
- Add new random parents:
  Adding random parents in certain optimization steps increases the chance to cover the whole design space. Although this method contradicts the idea of the optimization algorithms, it prevents the population from converging to local maxima.

2.1.3 Evolutionary vs. Genetic Strategy

Rolf Steinbuch

From the beginning of Evolutionary Optimization until now, there is a schism between two schools. The first of them, often called evolutionary direction, uses floating point representation of the parameters' values. Crossing is done by weighted averaging (cf. Fig. 2.1), where mutation means randomly changing values. The other direction, the Genetic Algorithms school, as it often labels itself, has a different view on the information stored in the genes of the parameter values representation. We all know that DNA which carries genetic information uses a

Fig. 2.4 From DNA to binary representation of crossing and mutation. (**a**) Original sequence of nucleobases. (**b**) Corresponding binary representation. (**c**) Crossing parents, dominating parents data printed bold. (**d**) Mutation of a binary child, mutated data printed bold

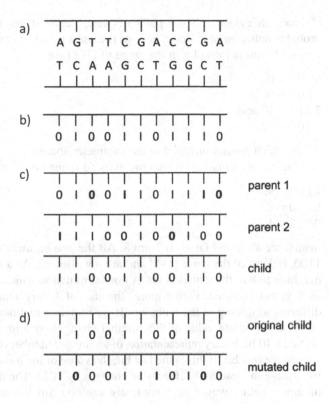

binary coding. Four nucleobases A, G, C, and T build ordered sequences in the helix (Fig. 2.4a). The nucleobase A matches with T, G, with C, and so a binary structure is defined. Copying is done by producing a negative of the original sequence and then refilling it with nucleobases of the original type. In the same manner we may look at the parameters' values in their binary representation (Fig. 2.4b). We are no longer interested in the numerical values they stand for, but only if the interesting binary digits (bits) are O's or I's. Combination means then to take one bit of the two parents (Fig. 2.4c), where mutation corresponds to a random change of single bits (Fig. 2.4d).

Generally there should be not a very large difference in the results of evolutionary or genetic processing. But in some cases, genetics may behave strangely.

Think of a parameter which takes values between 1 and 8. The binary representations of these values are as we all know

1. 0001
2. 0010
3. 0011
4. 0100
5. 0101
6. 0110
7. 0111
8. 1000

From an evidence based point of view, the next neighbor, and so the most probable mutation of 8, would be 7, as is closest to 8. But in a binary representation, the one bit mutations of 8 in the range of [1–8] are

$0000 = 0,$
$1001 = 9,$
$1010 = 10$ and,
$1100 = 12,$

all of which are not included in the parameter space.

The two bit mutations within the allowed range are

$0100 = 4,$
$0010 = 2,$
$0001 = 1,$

which are all rather far away from 8. All the one bit mutations 0000, 1001, 1010, 1100, fall out of the range [1–8] and are not allowed. As a result, there is no close neighbor to 8 in the mutation set. A one bit mutation which produces a value close to 8 is not possible. Furthermore, the use of binary representation provides a different weighting of the mutation. If we mutate a real number within a range of 10 %, the absolute value of this number changes by a maximum of 10 %. If we mutate a 10 bit binary representation of an integer number ($0 < x < 2047$) by 10 %, we change one bit. If this is the last bit, the value of the integer changes by 1, but if we change the leading bit, the value changes by 1024. The meaning of a maximum mutation radius would be completely useless. Small mutations in numbers of changed bits create essential changes of values, and thus the appearance of designs. We are somewhat skeptical of the use of binary parameter representation. Throughout this book, we deal with floating point values and fractions of them, when handling continuous parameters. Nevertheless, the genetic approach has its advantages, and in many applications there are no great differences in the final results of an optimization study using either school of thought. Chelouah and Siarry (2000) provide a detailed discussion of the reasons, why they and many others prefer genetic algorithms.

2.1.4 Discussion

One major advantage of Evolutionary Optimization is, in a design space with many local optima, it has the tendency to converge to the best solution if there are sufficiently large numbers of parents, kids, and generations, and if the mutation radius is rather large. This advantage is relative, as the number of individuals studied may become very large, even if the values driving the process are set in a favorable range.

2.2   Fern Optimization

Julian Pandtle

In addition to proliferation by crossing two parents, ferns and germane plants in nature also have the ability of asexual reproduction. This is the basic idea behind Fern Optimization. Unlike sexual reproduction, there is only one parent which generates kids without crossing its genes with other parents. The variation of the individuals' properties is only done by mutation (Smith et al. 2006).

2.2.1   Description of the Approach

A *number of parents* is chosen to reproduce themselves. Every parent generates discrete and completely independent *nkids* so that the difference between the kids and its parents varies only by the *mutation radius*. After that, the fitness of every kid is evaluated. From each parent the best kid is chosen to be parent of the next generation if it performs better than the parent (Fig. 2.5). These steps are repeated until either a convergence criteria is reached or a predefined number of generations have been calculated. The best individual of all kids of all initial parents over the sequence of generations is considered the optimum of the process. The pseudocode for Fern Optimization is shown in Table 2.2.

Periodically, in some generations we can check if the kids of the different initial parents fail to come up with relatively good results. We then remove from the total population those who seem to be not as fast in the current stage of the optimization process (Fig. 2.6). This accelerates the process essentially, as the number of

Fig. 2.5 Schematic diagram of Fern Optimization: one initial individual as parent, four kids, over three generations

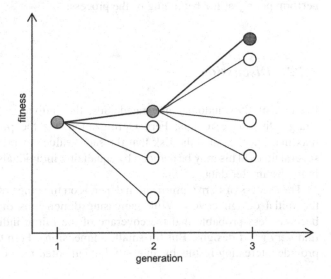

Table 2.2 Pseudocode: Fern Optimization

Initial: Define parameters:
  – Number of parents, kids per parent, generations
  – Mutation radius
  – Define *nparents* parents; evaluate their performance and restriction violation

Start *ngenerations* loop:
  – Define kids for each parent by mutation of its properties
  – Evaluate the kids' performance and restriction violation
  – Select the best kid of each parent (or keep the parent) for the next generation
  – After some generations, remove not promising families from the process

End loop

Stop

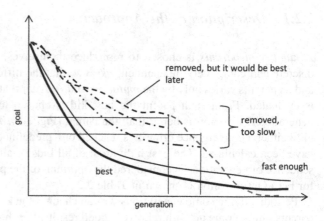

Fig. 2.6 Individuals offspring goals during a study—exclude slowly improving ferns

individuals to be handled significantly decreases. On the other hand, it may eliminate some good solutions that result from sequences of individuals that perform poorly at the beginning of the process.

2.2.2 Discussion

Because of the randomly generated kids, the individuals do not progress in a straight line. Due to this, Fern Optimization has the potential to ignore local maxima. There is also the risk that the same values or regions will be calculated several times. This may be avoided by equalizing individuals with small differences in the parameter data.

The success of Fern Optimization depends on the range of the solution space that the initial design covers. With increasing dimensions of the solution space, it becomes less probable that the coverage of the initial individuals is sufficient to find very good designs. But for smaller dimensions, Fern Optimization is able to provide interesting results by an easily implemented method.

There exist many variants of this Fern Optimization known e.g. as Tabu Search (Glover et al. 1999) or Neighborhood Search (Hansen and Mladenović 2005).

2.3 Particle Swarm Optimization

Simon Gekeler, Tatiana Popova, and Frank Schweickert

The Particle Swarm Optimization (PSO) (Kennedy and Eberhardt 1995; Plevris and Papadrakakis 2011) follows the observation that many groups of living beings have a tendency to behave similar to a complex being itself by local interaction and without centralized control. Well known examples are the swarms of fish or birds that seem to be orchestrated by a common intellect, known as Swarm Intelligence. The assumption is that they would not do so, if there was not an advantage compared to the free individual swimming or flying, which may be observed as well. We will not discuss here, what the advantage of coordinated behavior might be for these populations, but we try to explain how to translate it to an optimization strategy. Such as a swarm of fish looking for a food source in nature, in PSO a population of individuals drifts through the space of possible solutions, interacting as a swarm to find regions of good results. The basic assumption of PSO is that the individuals know their current position and velocity. Additionally, each individual remembers where its best position during the process has been, and it is informed where the best position in the parameter space found up to now is located. PSO is a global optimizer and able to solve constrained, non-linear problems in a multidimensional search space containing plenty of local optima.

2.3.1 Terms and Definitions

In PSO there is population of a user defined number of *individuals* with individual velocities, also referred to as *particles*, searching for possible optimization results in the search space.

In every *iteration*, all particles do one step in the search space.

Each of a particle's visited positions requires one objective function evaluation and represents one possible solution of the optimization process.

All of the particles know about their *personal best position* in the history of solution search.

In every iteration, the *global best position* is updated, which offers information about the best solution found previously by the whole swarm of particles.

The *particle's velocity* defines the individual direction and length of the particle's current step to adopt a new position in the search space. This velocity is updated every iteration by using the information of personal best position, global best position, and the particle's previous velocity.

As global settings to influence the behavior of the swarm's movement in the search space, we define the importance of the different contributions for the particle's velocity updates. The weight for the global best position is called the *social parameter*. The personal best position information is weighted by the *cognitive parameter*. And the particle's previous velocity is weighted by the *inertia parameter*.

2.3.2 Description of the Particle Swarm Optimization

Initialization of the Particle Swarm Optimization starts with randomly distributed particles within the upper and lower limits of the multidimensional search space. The position $\mathbf{p}_j$ of each particle j is given by the values of the free optimization parameters. In addition, for each particle an initial velocity $\mathbf{v}_j$ is induced. These velocity vectors are set up with random values, where the maximum possible value depends on the specific range of each optimization parameter. With the evaluation of the initial particles' fitness according to the objective function $z(\mathbf{p})$, the first global best position $\mathbf{p}_{Gb}$ is found. The particles' current position equals the primal personal best positions $\mathbf{p}_{Pb,j}$.

Afterward, in each iteration we update the velocity according to Eq. (2.1) for every particle j. Then the particles' new position $\mathbf{p}_j(t+1)$ is found by adding the new velocity $\mathbf{v}_j(t+1)$ to its current position $\mathbf{p}_j(t)$ (cf. Eq. 2.2). Before starting the next iteration, the fitness of all particles is evaluated and the global best and all personal best positions are updated. This loop is continued until a stop criterion e.g. the maximum number of iterations is reached.

$$\mathbf{v}_j(t+1) = c_1 \mathbf{v}_j(t) + c_2 \mathbf{r}_1 \circ \left(\mathbf{p}_{Pb,j} - \mathbf{p}_j(t) \right) + c_3 \mathbf{r}_2 \circ \left(\mathbf{p}_{Gb} - \mathbf{p}_j(t) \right) \qquad (2.1)$$

$$\mathbf{p}_j(t+1) = \mathbf{p}_j(t) + \mathbf{v}_j(t+1) \qquad (2.2)$$

Here, the inertia parameter c_1 weights the previous particle velocity $\mathbf{v}_j(t)$, the inertia tendency, forcing the particle to explore the search space by continuing its direction of travel. The cognitive tendency makes the particle try to return to its individual best position $\mathbf{p}_{Pb,j}$ discovered before. It is weighted by the cognitive parameter c_2 and influenced by the random vector $\mathbf{r}_1$, with entries uniformly distributed in the interval $[0, 1]$. The social parameter c_3 and the random vector $\mathbf{r}_2$ weight the social tendency, which pushes the particle to try to reach the best position found so far and forces the particle swarm to converge. The product $\mathbf{r} \circ \mathbf{p}$ is the vector produced by elementwise multiplication, the Hadamard product. The principle scheme of updating a particle's velocity in Eq. (2.1) is visualized in Fig. 2.7.

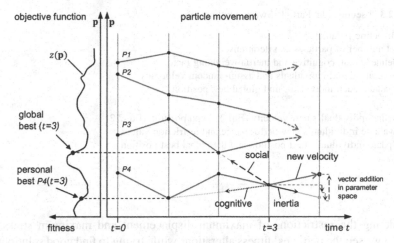

Fig. 2.7 Definition of the particles velocity components

To define the weighting factors c_1, c_2 and c_3, there is no rule generally, as it depends on the optimization problem and the available computation time to find a result. The interaction of all parameters is important. We can either force the particle swarm to converge and obtain optimization results relatively fast or we can attach importance to the exploration of the search space, trying to find the global optimum with the cost of additional variants needed to be computed. Higher values of the inertia parameter c_1 cause a long exploration phase before the swarm converges, likely to the global best position. However, more weighting by the social parameter c_3 leads to an earlier convergence. A limit value of the inertia parameter $c_1 < 0.95$ is important, as otherwise the swarm will not converge and PSO will behave similarly to a random search in the search space. Also, the social and cognitive parameters should follow the rule $0 < (c_2 + c_3) < 4$, as otherwise the particles would oscillate around the global best position again without converging (Perez and Behdinan 2007).

To handle constrained optimization problems, restrictions may be managed with the penalty method (cf. Sect. 2.9).

A pseudocode of a simple PSO realization is shown in Table 2.3. The updating of global and personal best information in the sequence of the particle swarm algorithm might be done in two ways. If we update during the iteration loop, each time a particle's new position is evaluated, the swarm communicates quickly and next particles are able to react directly to new global best positions. The other method is to update the coordinates only once, at the end of each iteration. This offers the possibility to parallelize the computation of particles (cf. Sect. 3.3.2).

Example 2.1 As an example of the PSO process, we optimize the well-known benchmark problem 10 rods frame (cf. Sect. 3.1.1, Fig. 3.1b). To minimize the frame's mass, we use 15 particles and 40 iteration steps. With the weighting factors of $c_1 = 0.6$, $c_2 = 1.3$ and $c_3 = 1.7$, the PSO is able to find the global optimum while

Table 2.3 Pseudocode: Particle Swarm Optimization

Initial: Define parameters: – Set number of particles and iterations – Define social, cognitive and inertia weighting factor – Generate initial individuals and assign random velocities – Evaluate each individual; find global best position
Start loop: – Define individual's new velocity (Eq. 2.1) and position (Eq. 2.2) – Evaluate individual's new performance and restriction values – Update individual's best position and the global best position
End loop
Stop

considering the restrictions of maximum displacement and maximum stress. In Fig. 2.8 we see the particles' fitness alterations while trying to find good solutions in the search space. The reason for excessive high peaks, especially at the end of the process, is the punishment of fitness values by the penalty method when particles exceed the stress or displacement limit. In addition, the convergence behavior of the best solution, found in the particle swarm process, is plotted. It indicates that if we want to use a stop criterion from the convergence of the best solution, the convergence rate must be observed over several iteration steps before the algorithm is stopped, as there could be a significant improvement after a period (e.g. iteration 4–7) without any advance.

Variants

Since the first version of the PSO, developed by Kennedy and Eberhardt (1995), was released, many different variants (Dorigo et al. 2008), hybrid procedures (Plevris and Papadrakakis 2011) and modifications have arisen. Most of them are adaptions to specific problems. One basic approach is the PSO procedure we have been introducing. It adds the inertia weight coefficient to the former version of the velocity update (Shi and Eberhart 1998). This enables further control of the optimum search. Based on this approach there are multiple ways for the particles' communication. We often use the global best neighborhood topology (Fig. 2.9a), in which each particle is informed of the best position in the whole population so far. The global best topology is a fast and reliable method for diverse optimization problems, but there are also different local neighborhood topologies, which represent slower communications and thus force the exploration of the search space, usually with higher computational costs (Kennedy 1999). The communication of the particles is limited either to a specific geographic neighborhood region in the search space or by a predefined particle linkage, as visualized in Fig. 2.9b, c.

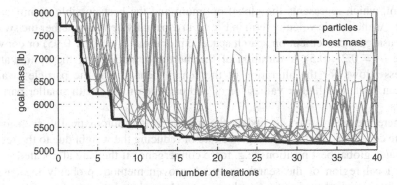

Fig. 2.8 Convergence of the global best solution and fitness of the particles positions when minimizing the mass of the 10 rods frame (cf. Sect. 3.1.3, Fig. 3.1b)

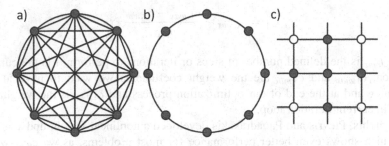

Fig. 2.9 Examples of neighborhood topologies for the particles communication in PSO. (a) Global best topology. (b) Local ring topology. (c) Local von Neumann topology

Other basic PSO procedures:

- The *Bare-Bones Particle Swarm Optimization* (BBPSO), where the update of a particle's velocity is replaced by an update of the particle's position following a probability distribution (Kennedy 2003).
- For operations in discrete spaces, there is the *Binary Particle Swarm Optimization* (Kennedy et al. 1997).
- The *Fully Informed Particle Swarm* (FIPS), where a particle's velocity update follows the information provided by all of its neighbors (Mendes et al. 2004).

In addition, the performance of the introduced PSO procedure may be increased by implementing a dynamic inertia weight update or by limiting the particles' velocity. These modifications will be explained more in detail in the following sections.

2.3.3 Dynamic Particle Inertia

At first, when doing an optimization with many local optima, we need to find the highest hill or deepest valley in the search space and then reach its peak or the bottom, which represents the global optimum and the best possible optimization result. As the inertia velocity term in Eq. (2.1) especially defines whether the swarm is focusing its movement on search space exploration ($c_1 = 0.5 \ldots 0.95$) or convergence ($c_1 = 0.2 \ldots 0.5$), a variation of the inertia weight during the optimization process allows for the alteration of the search behavior of the particle swarm. We can start with higher values for c_1 and end the process with smaller particle inertia.

There are many ways to update the inertia weight automatically, such as an update dependend on the convergence rate or reducing the weight due to the recent evaluated global best positions, e.g. force convergence if they are all located in the same local region of the search space. A proven method, probably the easiest implementation of an inertia weight update, depends on the number of iteration steps we want to fulfill in our optimization. E.g. the implementation of a linear decrease (Eq. 2.3) with the ongoing process time (Shi and Eberhart 1998).

$$c_{1,t+1} = c_{1,max} - \frac{c_{1,max} - c_{1,min}}{t_{max}} * t, \qquad (2.3)$$

where t_{max} is the defined number of steps or iterations to perform, t is the current iteration, $c_{1,max}$ and $c_{1,min}$ are the weight coefficients we want to have at the beginning and at the end of our optimization process and $c_{1,t+1}$ is the weight we apply in current iteration loop.

From this, Plevris and Papadrakakis developed a nonlinear weight update strategy which shows even better performance for most problems, as we can switch more quickly to the convergence phase (Plevris and Papadrakakis 2011). We use a cubic interpolation of the 4 given points, shown in Fig. 2.10a, to obtain the inertia weight update function. The calculation of coefficient b is

$$b + a_w * b + a_w^2 * b = c_{1,max} - c_{1,min}$$
$$\Rightarrow b = \frac{c_{1,max} - c_{1,min}}{a_w^2 + a_w + 1}.$$

Here a_w influences the shape of the update function, as indicated in Fig. 2.10b.

We should keep in mind that a quick convergence in a population based optimization does not guarantee the finding of the global optimum. A disadvantage with the time dependent inertia updates occurs when we want to use a stop criterion dependent on the convergence rate. It is evident that, if the particle swarm has already found a good solution in the exploration phase, it is not recommended to stop the algorithm. More iterations are required to further reduce the inertia parameter to gain additional improvement in the swarms convergence phase.

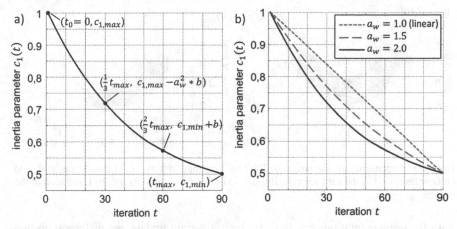

Fig. 2.10 Nonlinear dynamic inertia weight update in PSO according to Plevris and Papadrakakis (2011). (**a**) Point coordinates, here e.g. $t_{max} = 90$, $c_{1,max} = 1$, $c_{1,min} = 0.5$ and $a_w = 2.0$. (**b**) Influence of variable a_w on the update function

2.3.4 Limitation of the Particles' Velocity

Misfit settings of the velocity weights lead to poor results of the PSO process as the swarm movement grows out of control. If the particle steps are too large they may jump over promising regions in the search space, so the process behaves similarly to a random method. Limiting the velocity can help to make such a PSO more reliable. Of course, too strong velocity limitation will restrict the method's exploration ability. Using a half-width of the optimization parameters' range should be not a bad choice (Gekeler 2012).

Figure 2.11 compares the results found for the 13 rods frame (cf. Sect. 3.1.3, Fig. 3.1c) for different weighting factors of the inertia and the social component of the new velocity. We see a valley of efficient combinations of the velocity weighting factors. This valley is limited by steep hills indicating less efficient progress and weaker goals achieved by the PSO studies. Limiting the particle's velocity to a certain maximum value (Fig. 2.11b) has the potential to reduce the height of the valley's side walls, but reduces the convergence of the optimization process (Gekeler et al. 2014).

2.3.5 Discussion

PSO has proven to be very successful if an appropriate set of particles and velocity-weighting factors $\{c_1, c_2, c_3\}$ has been selected. This method has the tendency either to stick to local optima or to perform similarly to a random search if these parameters are not well chosen (cf. Fig. 2.11) (Gekeler et al. 2014). Especially

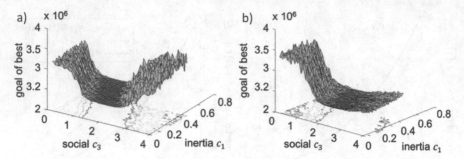

Fig. 2.11 Dependency of the power of PSO of weighting factors for 13 rodsframe, with and without limitation of the particles velocity. (**a**) No limitation of the velocity. (**b**) Limitation of the velocity

with the implementation of a dynamic inertia weight update, we differentiate two stages of the PSO process, first the exploration phase and then the following convergence phase. The weighting factors help the user either to focus the search on the exploration of the search space or on fast convergence of the optimization process. The latter forces a quick search to find satisfying solutions, but is probably not the best choice, because of limited exploration in the search space. Hence, getting experience of choosing proper PSO settings for the different optimization problems is necessary. Section 3.3.1 offers proposals for the adequate definition of the PSO parameters for the first trials of this optimization procedure. Please remember, that PSO is a population based optimization method and the number of function evaluations may be large.

2.4 Artificial Neural Net Optimization

Christoph Widmann

The human brain is one of the most complex of known structures and superior to modern computers in its cognitive abilities. It consists of more than 100 billion neurons which are connected by a net of up to 100 trillion links. To emulate and exploit this biological organ for technological purposes, scientists have built artificial models that copy the structure of the brain. These so called artificial neural nets (ANNs) are simplifications of the complex brain structure.

In optimization, ANNs are used to represent the behavior of the whole optimization problem in one neural net. Unlike Response Surfaces, (see Sect. 2.7), it is not necessary to know anything about the dependencies of the design variables. ANNs are able to learn those complex dependencies without knowing them at the beginning. What has to be done at the beginning is to define the architecture of the ANN.

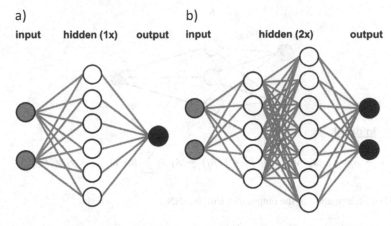

Fig. 2.12 ANNs with two different networks. (**a**) Network: 2-6-1, three layers. (**b**) Network: 2-5-7-2, four layers

2.4.1 ANN Architecture

In contrast to the biological brain, ANNs are built up in different layers (Fig. 2.12). The layers are divided into:

- the **input layer** where neurons receive signals from the environment
- one or more **hidden layer** which process the output signals from the input layer and transmit them to the next hidden or output layer
- the **output layer** which is fed by the (last) hidden layer. The outputs of these neurons represent the response of an input to the ANN.

Figure 2.12a shows an ANN with two input neurons, one hidden layer with six neurons and one output neuron. These types of ANNs are also called feed forward nets because the information is only transferred in one direction. In most applications of ANNs, only one or two hidden layers (Fig. 2.12b) are used, as they suffice to represent even very non-regular outputs.

Every neuron of a layer is connected with all the neurons of the following and preceding layer (Fig. 2.13). The special feature of the connection between the neurons is that every signal, e.g. x_i, is weighted by a specific weight factor $w_{1,ij}$. With this factor, the signal is increased or reduced. In these weighting factors, the whole knowledge of the ANN is stored.

Like the biological model of neural nets, every neuron (except neurons of the input layer) needs an activation potential. Every incoming (weighted) signal to a neuron is summed up to a value, e.g. A_j. With a special transfer function $T(A_j)$, the value of the neuron, e.g. h_j is calculated. There are many different transfer functions known and discussed by, e.g. Lagaros (Lagaros and Papadrakakis 2004), but in most ANNs the sigmoid function or a variant of it is being used:

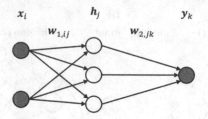

In detail:

$$x_i \rightarrow A_j = \sum_i x_i * w_{1,ij} \quad \rightarrow \quad h_j = T(A_j) \quad \rightarrow \quad A_k = \sum_j h_j * w_{2,jk} \quad \rightarrow \quad y_k = T(A_k)$$

Fig. 2.13 Determination of the output of a simple ANN

$$T(A) = sig(A) = \frac{1}{1 + e^{-A}} \tag{2.4}$$

Many more ANN architectures are known and in some cases it can be necessary to use a different net architecture see Zell (1997) and Günther et al. (2010).

2.4.2 Training ANNs

There are many different methods to train an ANN. All of these methods follow an optimization process themselves. The goal of this optimization is to minimize the squared error E (c.f. Eq. 2.5) between a training set of known input and expected output values and the output generated by the ANN. This is done by modifying the weighting factors. This most commonly used training method is called supervised learning, where the number of necessary training sets is dependent on the architecture and the complexity of the optimization problem.

$$E = \frac{1}{2} \sum_p \sum_k \left(d_{kp} - y_{kp} \right)^2 \tag{2.5}$$

The error of one training pair is calculated by the squared difference of the net output y_{kp} and the expected output d_{kp} were k is the index of the output neuron and p is the index of the training pair. The total error is then the sum of training pair errors E. To measure the training progress or the convergence of the ANN, it is important to define a testing set as well. The error of the testing set is calculated analogously to the training set error. In some cases it can be possible to over-train an ANN, e.g. the representation of the problem becomes worse with ongoing training when using an insufficient number of training sets or the architecture of the net cannot handle the complexity of the problem. This can be indicated by steadily

Fig. 2.14 2D-Schwefel function used to be interpolated by an ANN

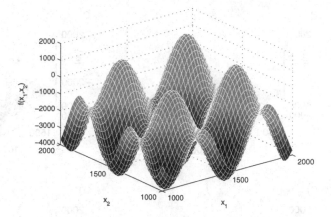

increasing error values for the testing set. In that case, it may be wise to save the actual weight factors of the net.

Most of the many training techniques are based on gradient methods for the weighting optimization. These methods include the Hebb-rule (Hebb 1949), Delta-rule (Widrow and Hoff 1960) and the backpropagation algorithm. One optimized backpropagation algorithm is the Quickprop algorithm by Scott Fahlman (Fahlman and Lebiere 1991), who used a quadratic Gradient Method for the weight optimization. A nonlinear least square training method was proposed by Levenberg and Marquardt (Hagan and Menhaj 1994). Compared to the previously mentioned methods, the change in weights is done after a run throughout the whole training set. It is also possible to use any other optimization technique to optimize the weight factors, for example evolutionary algorithms.

Example 2.2 In the following example, an ANN was trained to approximate a part of the Schwefel function (Fig. 2.14). The Schwefel function (Schwefel 1989) is given by:

$$f(x) = 418.9829 * d - \sum_{i=1}^{d} x_i * \sin\left(\sqrt{|x_i|}\right), \qquad (2.6)$$

where d is the dimension of the problem. In this example we use $d = 2$ and the considered range within $x_i \in [1000, 2000]$.

To begin, a set of design points has to be chosen. The number of designs depends on the complexity of the problem and should cover the whole area of interest. In our shown example, there are 100 designs distributed by a Latin Hypercube sampling.

The next step is to scale the input and output values (design pairs) in a way that they are between $[0...1]$. This is a common use, needed for most ANN training algorithms. If the design space varies in different dimensions too much, e.g. x_1 $[1...10]$ and x_2 $[100...1 * e^5]$, it might be necessary to scale each dimension separately so that that there is no loss of information.

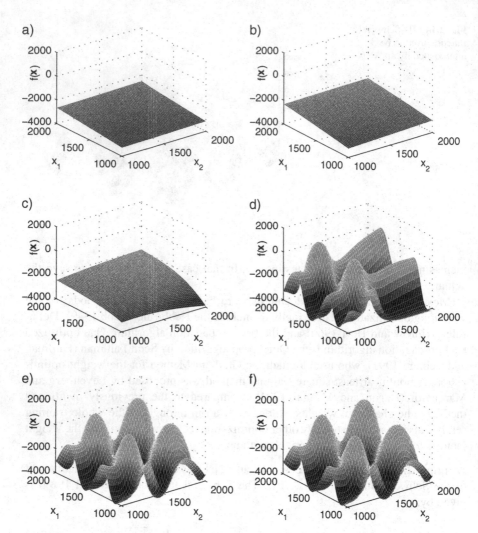

Fig. 2.15 Approximation of the Schwefel function with an ANN during its learning progress. (**a**) After 10 runs learning. (**b**) After 100 runs learning. (**c**) After 1000 runs learning. (**d**) After 10,000 runs learning. (**e**) After 50,000 runs learning. (**f**) After 100,000 runs learning

For our example of an ANN architecture, we used 20 neurons in one hidden layer. It was trained with a standard backpropagation algorithm. Figure 2.15 shows the ANN approximating the Schwefel function at different learning stages after 10, 100, 1000, 10,000, 50,000 and 100,000 training runs.

It can be seen that during the course of ANN training the approximation becomes more sensitive to the training data. At the beginning, there was a nearly linear approximation, later a more or less quadratic one, and at the end a complex dynamic approximation. To measure the quality of the ANN approximation, the coefficient of determination of the learning and test examples can be calculated. After training the ANN, an optimization on the approximated surface might be done with one of the algorithms discussed in this book.

2.4.3 Conclusion

The amount of computation to qualify a neural net may be large, but it is a powerful tool if there are many function evaluations to be done, and a single evaluation requires much computing power itself. The ANN acts as a meta-model (cf. Sect. 2.7) which is easily capable of producing the required output with a known exactness. Based on this meta-model, an optimization might be done.

2.5 Ant Colony Optimization

Tatiana Popova

In a similar manner to swarms of birds or fishes, from which we got the idea for the Particle Swarm Algorithm (see Sect. 2.3), ants live in very interesting social hierarchies that exhibit complex behaviors. Ants have been attracting human attention for many years. Ants are typical examples of the phenomena called Swarm Intelligence. An ant colony consists of a set of individuals, the ants, with relatively simple responses to their environment. The ants are not underneath global control: they do not obey direct orders from a central command. Rather, they organize themselves by communication between individuals within their colony. Self-organization is provided by direct communication (visual or chemical contact) or indirect communication (stigmergy) between the colony's members. Using such simple self-organization, even complex tasks can be accomplished.

In their colonies (ants, bees, etc.) many tasks need to be fulfilled. Some of them are brood tending, foraging for resources or maintaining the nest. Division of labor is the main challenge. The coordination of those tasks requires a dynamic allocation of individuals to different tasks. This coordination may depend on the state of the environment or needs of the colony. Consequently, a global assessment of the colony's current state is required to most effectively assign tasks to individuals. But the individuals themselves are unable to make a global assessment. The ants' solution seems to be the use of response threshold models.

An individual labelled i engages in task j with a probability that depends on the structure of a stimulus:

$$p_{ij} = \frac{s_j^2}{s_j^2 + \delta_{ij}^2} \qquad (2.7)$$

In Eq. (2.7), each individual i has an individual response threshold δ_{ij} for each task j and $s_j \geq 0$ the stimulus of task j. When the stimulus s_j is much smaller than the response threshold δ_{ij} the probability p_{ij} to engage in that task is close to 0. If the stimulus s_j is essentially greater than the response threshold δ_{ij} the probability is close to 1. With the proper stimulus, that ant will tackle that task. In such a manner, labor division is organized within a colony.

2.5.1   The Ant Colony Strategy in Bionic Optimization

The most remarkable behavior for us is the formation of "ant streets." According to many researchers who have been studying ants in detail, one of their most surprising abilities is to find the shortest path (Dorigo and Stützle 2004; Stock et al. 2012; Kramer 2009). It has been shown that the ants use communication based on pheromones that they may deposit and smell. This behavioral pattern inspired computer scientists to develop algorithms for the solution of optimization problems. The first attempts at applying this method appeared in the early 1990s, but had no practical applications.

In the real world, ants ramble randomly, and upon finding food, return to their colony. They deposit pheromone trails constantly marking the paths they have chosen. If another ant finds such a path, it tends not to continue travelling at random, but instead follows the trail, returning and reinforcing it if it finds food. The importance of the trail is improved by additional ants travelling along it and depositing yet more pheromones.

An example, improving the performance of foraging, may explain the importance of this street definition for following ants. To organize foraging by transporting food to the nest, depicted in Fig. 2.16, ants coordinate their activities via stigmergy. This stigmergy is a mechanism of indirect coordination between agents, here the ants, or actions mediated by modifications of the environment. A foraging ant deposits a chemical, the previously mentioned pheromone, on the ground which increases the probability that other ants will follow the same path (Dorigo and Stützle 2004).

2.5.2   Description of the Approach

Let's imagine the situation shown in Fig. 2.16. Ants are free to move between the nest and the food source. There are two ways to the food source: one long and one short. Here, the short one is clearly more efficient for foraging, and needs to be

Fig. 2.16 Biological ants
order

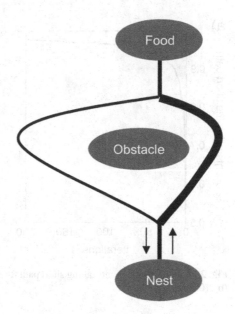

found. For every path, we identify τ_1, τ_2-artificial pheromone parameters. Each
indicates the expected benefit when choosing that path. As initially
$\tau_1 = \tau_2 = c > 0$, it is necessary to provide the same choice probability for both
paths at the beginning.

In the first iteration, every ant chooses one path according to its probability of
benefit for reaching the food source and then getting back to the nest. Thereby each
ant leaves pheromone on the traversed edge. In the algorithm, this requires an
update of the pheromone parameters according to Eq. (2.10), after the first ants have
returned to their nest and the covered distance l_i is known.

After the initial iteration, this probability is calculated according to the formulas:

$$p_1 = \frac{\tau_1}{\tau_1 + \tau_2}, \tag{2.8}$$

$$p_2 = \frac{\tau_2}{\tau_1 + \tau_2}, \tag{2.9}$$

where

$$\tau_{i,new} = \tau_{i,old} + \frac{1}{l_i}. \tag{2.10}$$

In consequence, the probability that the ants will choose the shorter path in the
following iteration increases as the probability is higher due to the update of
pheromone parameters in Eq. (2.10). Additionally, the trail of old pheromones
evaporates and non-promising paths will lose the chance to be chosen.

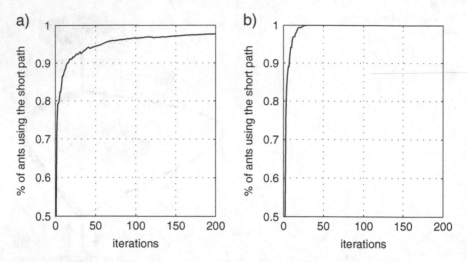

Fig. 2.17 Percentage of ants taking short path for different population size. (**a**) 10 ants population. (**b**) 100 ants population

After some time or multiple iterations, all the ants chose to use only the short branch with the stronger signal. For this simple example, Fig. 2.17 depicts the percentage of ants taking the short path, for population of 10 and 100 ants. We can see that an algorithm with 100 ants is essentially faster, but 10 ants are able to converge to the shortest path as well.

It has been shown that it is often sufficient to consider a stigmergic, indirect mode of communication to explain how social insects can achieve self-organization. Ant algorithms use artificial stigmergy to coordinate societies of artificial agents. An ant colony algorithm is a probabilistic technique for solving computational problems that can be reduced to find optimal paths through graphs.

Ant Colony Optimization (ACO) algorithms have been tested on a large number of academic problems. These include traveling salesman problems (Fig. 2.18), where the goal is to find the shortest route that visits all given places only once, as well as assignment, scheduling, subset, and constraint satisfaction problems. This success with academic problems has raised the attention of industrial users who have started to use ACO algorithms for real world applications. For instance ACO is applied to a number of various scheduling problems such as a continuous two-stage flow shop problem with finite reservoirs or vehicle routing problems.

ACO algorithms are very successful and widely-recognized algorithmic techniques based on ant behaviors. Their success is evidenced by the extensive array of various different problems to which they have been applied, and moreover by the fact that ACO algorithms are currently, for many problems, among the top-performing algorithms. They could be used for design optimization as well, but we have learned from experience that they do not perform too well, so we do not extend the discussion here.

Fig. 2.18 Traveling salesman problem: find shortest route to visit all locations

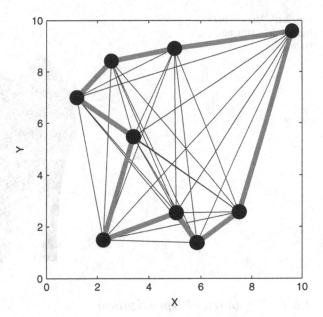

2.6 Non-parametric Optimization

Stephan Brieger

A typical optimization problem in structural design is to find the optimal layout or shape of a structure within a specific region. Often the only known quantities are loads, supports, and constraints such as mass or material restrictions. With this starting point, the problem is not easily represented by design parameters. The purpose of non-parametric optimization is to give engineers a method to define a design space in regions or whole components without the process of defining the problem in design parameters. Frequently used non-parametric structural optimization methods are Topological and Local Growth Optimization (Bendsoe and Sigmund 2003; Haftka and Gürdal 1992; Rozvany 1997; Vanderplaats 1984). In the early stage of concept generation, Topological Optimization can be used to develop an efficient structural layout. In a later process of product development, Local Growth Optimization is an efficient tool to fine-tune the optimized structural proposal.

In the process of these optimization methods, the software implementation automatically does a parameterization of the design space. Standard mathematical optimization techniques are often not suitable for these problem formulations due to the high number internal design variables.

Fig. 2.19 Human bone—a
light-weight structure
(femur)

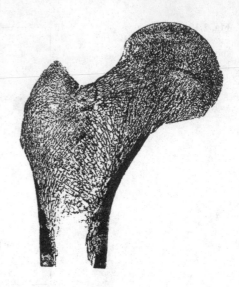

2.6.1 Topological Optimization

Topological Optimization is a method for optimizing material layout within a
defined design space with respect to loads and constraints. The process of Topological Optimization is similar to the process of bone mineralization in living
creatures (Fig. 2.19). This biological process leads to a stiffness-optimized structure
with minimum stresses and minimum weight by modifying the material distribution
towards highly loaded areas.

In general the implementation consists of a Finite Element Analysis combined
with an optimization technique for iterative updates to the material distribution. The
design space is divided into small regions of varying density. Here we often use
Finite Elements to define these regions. To find an optimized design, the density of
each element of the FE-meshed design space is adjusted by an optimizer to match
desired objective and constraints. No shifting of nodes is performed. In practice, the
material density has been adjusted by modifying the stiffness of the corresponding
Finite Element.

The result of a Topological Optimization run is a density field of the design
space, which needs to be interpreted in most cases. Domains with high stiffness
shape the structural design proposal of the component; domains with low stiffness
form void areas.

In general the resulting design proposal needs to be re-designed and fine-tuned to
satisfy manufacturing requirements. The main steps of Topological Optimization
are represented in Fig. 2.20. The corresponding pseudocode is shown in Table 2.4.

Structural proposals by Topological Optimization often are infeasible to manufacture. Because of this, most commercial CAE-Software for Topological Optimization is commonly extended with features for considering manufacturing

Fig. 2.20 Topological
Optimization: from problem
definition to structural
design proposal.
(**a**) Problem definition.
(**b**) Result of Topological
Optimization.
(**c**) Re-designed proposal
to satisfy manufacturing
requirements

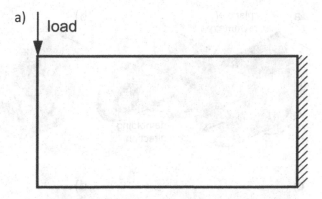

Table 2.4 Pseudocode:
Topological Optimization

Initial: Define problem and parameters:
  – Define design space
  – Define boundary conditions

Start *ngenerations* loop:
  – Evaluate FE-model
  – Adjust density(and stiffness) of each FE-element

End loop
Stop Derive structural design from proposal

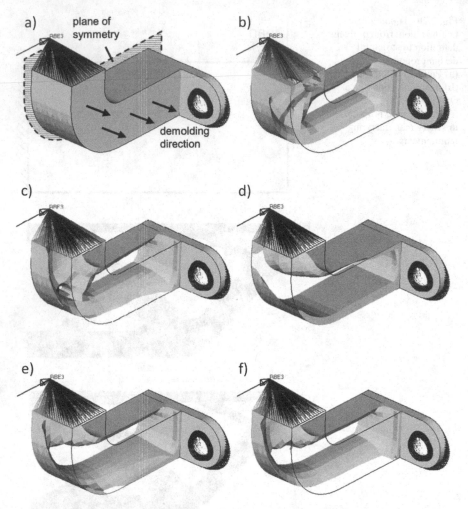

Fig. 2.21 History of Topological Optimization of a trailer coupling, taking into account the direction in which the casting tools are removed. (**a**) Given design space at t_0. (**b**) Design proposal at t_1. (**c**) Design proposal at t_2. (**d**) Design proposal at t_3. (**e**) Design proposal at t_4. (**f**) Final design proposal at t_{end}

constraints, such as casting constraints (Fig. 2.21), symmetry and pattern constraints, or member sizing directly in the formulation of the optimization problem.

The fact that the user does not need to define complex design variables makes Topological Optimization an excellent tool in the early stages of design processes to find a first design proposal or to understand basic load paths.

2.6.2   Local Growth

Biological structures such as tree stems change their own shape by growth and shrink their surfaces to adapt to external loads (c.f. Fig. 1.3b). Stress peaks are reduced by adding material at high stressed surface areas. The volume in low stressed areas is reduced in size by shrinking the surface. This biological growth rule creates light-weight structures with minimized notch stresses and maximized stiffness. Since this is often the preferred objective in structural optimization problems, the local surface growth process is adapted as an optimization tool in many CAE-Systems. There are various different approaches to solving these problems. With simple optimality criteria methods or growth rules, one can get good results.

The process of simulated growth is based on an iterative process of FE-simulations and an optimization technique that updates the surface to change the shape of the structure to meet with objectives and constraints. The shape perturbations are either manually defined by the user or automatically determined by the CAE-System. A common way to describe the shape changes of the Finite Element model is to define some shapes as a perturbation $\mathbf{b}$ of nodal coordinates $\mathbf{r}_0$.

$$\mathbf{r} = \mathbf{r}_0 + \mathbf{b} \qquad (2.11)$$

The new design can be generated by doing a linear combination of these shape vectors. The design variable is defined as the weighting factors w_i of the shape vector,

$$\mathbf{r} = \mathbf{r}_0 + \sum_{i=0}^{n} w_i \mathbf{b}_i \qquad (2.12)$$

where n is the number of shapes and design variables.

Figure 2.22 shows an example for optimizing notch stresses. We start with an initial geometry. The defined region we want to optimize is given by a simple shape

Fig. 2.22 Notch stress optimization results in smooth change of shape. (**a**) Problem definition. (**b**) Optimized shape

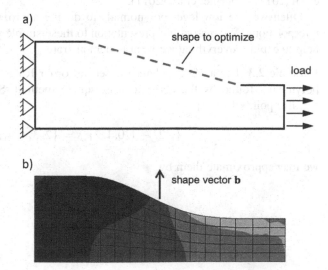

representation. After some iterations, we get a smooth shape with minimized notch stress that takes into consideration the applied loads and constraints.

2.7   Meta Models

Rolf Steinbuch

When we do any optimization or sensitivity study, we produce sets of data of goals, and restrictions and other information in a certain region of the space of possible or acceptable parameters. As the computation of each of these variants may be costly in terms of time and computing power, we generally should restrict the number of variants to the absolute minimum. From this motivation came the idea of meta-models. *Surrogate models* such as quadratic Response Surfaces (RS), which we will discuss in more detail later, and polynomial chaos expansions or Kriging (which are built from a limited number of runs of the original model) have been introduced as substitutes for the original time consuming FE-job model to reduce the total computational cost.

As designers in mechanical engineering, we are dealing with variants of a basic design of a macroscopic component. As the changes of the parameters are limited, the change of the systems response is limited and more or less smooth as well, when examined as functions of their parameters. So why don't we use a sufficient number of data in an interesting region and approximate the responses by simple and smooth functions? If there are no catastrophic changes in the results of the computation, their representation, often called Response Surface (RS), should be a fairly regular function. Such functions often may be approximated by smooth functions. There are many other ways to define meta-models see, e.g. McKay et al. (1979), Au and Beck (1999), Das and Zheng (2000), Matthies et al. (2013), Dubourg et al. (2013), Bourinet et al. (2011).

Often we use low-level polynomials to do the approximation, e.g. parabolic interpolations. We restrict our presentation to these simple polynomial RS, as they help to explain everything we want to demonstrate.

Example 2.3 For a 1D Problem, the second order RS is nothing more than the polynomial found by the classical least square method. So if we take a set of, e.g. five points $\{(p, s)\}$

$$p = \{-2, -1, 0, 1, 2\}; s = \{2, 3, 4, 4, 2\}$$

we may approximate them by

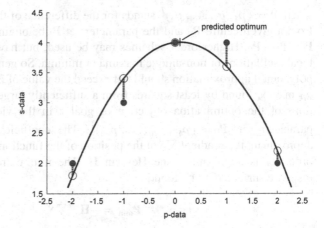

Fig. 2.23 1D Response Surface used to find the optimum of a set of data points

$$s = -0.5p^2 + 0.1p + 4, \qquad (2.13)$$

a parabola which is plotted together with the data points in Fig. 2.23. The squared averaged error (Eq. 2.18) of this approximation would be

$$error = 0.368.$$

If this error is small enough, we may use the approximation to do local studies. For example, we can search for the position of the local maximum of this curve, which is indicated in Fig. 2.23 as well, or may be found by

$$s' = -p + 0.1 = 0 \qquad (2.14)$$

to be at the p-value

$$p_{max} = 0.1.$$

In the general case of a n-dimensional problem, RSs are defined by low-order polynomial functions of the free parameters as well. For a second order approximation, we may use the notation

$$
\begin{aligned}
S &= a_0 + a_1 p_1 + a_2 p_2 + \ldots + a_n p_n \\
&\quad + a_{11} p_1 p_1 + a_{12} p_1 p_2 + \ldots + a_{n-1,n} p_{n-1} p_n + a_{nn} p_n p_n \\
&= a_0 + \sum_{k=1}^{n} a_k p_k + \sum_{k=1}^{n}\sum_{l=k}^{n} a_{kl} p_k p_l \\
&= a_0 + \mathbf{P}^T \nabla S + \frac{1}{2}\mathbf{P}^T \mathbf{H} \mathbf{P},
\end{aligned}
$$

$$(2.15)$$

where $\mathbf{P} = (p_1, p_2, \ldots, p_n)^T$ stands for the difference of the centred parameter set from a given centre $\mathbf{P}_0$ and the parameter set $\mathbf{P}_i$ belonging to the i-th evaluation $\mathbf{P} = \mathbf{P}_i - \mathbf{P}_0$. Higher order schemes may be used, but have the tendency to show local oscillations or non-unique maxima or minima. So generally, the degree of the polynomial approximation should not exceed the value of 2. The coefficients a_i and a_{ik} may be found by least squares from a sufficiently large set of function evaluations of the optimization objective or goal z in the vicinity of an interesting parameter set $\mathbf{P}_0 = (p_{10}, p_{20}, \ldots, p_{n0})^T$. The coefficients of the linear terms approximate the gradient ∇S at the position of the function evaluation. The second order terms approximate the Hessian $\mathbf{H}$. The minimum (or maximum) of the response surface may be found by

$$\mathbf{P}_{\min} = -\mathbf{H}^{-1} \nabla S, \tag{2.16}$$

which is often a good guess for the position of the exact optimum of $z(p_1, p_2, \ldots, p_n)$ as long as the function does not behave too irregularly. One sign that this is regular is that the determinant $\mathbf{H}$ must not be singular. To be more specific, $\mathbf{H}$ should be either positive or negative definite, meaning its Eigen values are all positive or all negative. To determine the coefficients a_i and a_{ik}, more than

$$n_{coeff} = 1 + n + \frac{n(n+1)}{2} \tag{2.17}$$

function evaluations are necessary, a number which may be rather large, especially in the case of an optimization with many free parameters.

Both, gradient method (cf. Sect. 1.4) and Response Surface approximations often help to locate an optimum faster than the bionic methods which locally do a random search and tend to miss the exact solution.

There should be no doubt that this interpolation is only meaningful if there are essentially more data points available than the coefficients we are trying to find. Figure 2.24a shows an interpolated RS and the data points used to generate the coefficients for a 2D problem. The quality of this surrogate may be again checked by summing up the maximum differences between RS and input data f.

$$error = \frac{1}{ndata} \sum_{i=1}^{ndata} (RS(\mathbf{P}_i) - f(\mathbf{P}_i))^2 \tag{2.18}$$

If this error is small enough, we may use the RS instead of performing more function evaluations, e.g. by time consuming FEM-Jobs. As in the 1D case, the optimum may be found by a solution using Eq. (2.16) or by an iterated search such as indicated in Fig. 2.24b.

In Sect. 2.8 we shall return to this example to apply the response surface to a numerical integration.

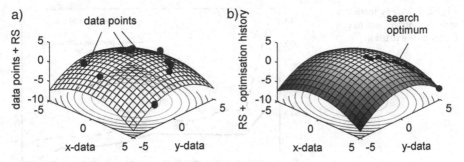

Fig. 2.24 2D-RS set-up and use to find a local optimum. (**a**) Data points and RS generated. (**b**) RS used to find optimum

As the calculation of the coefficients in Eq. (2.15) does not take much time compared to a finite element job, it is a good idea to determine the RS as soon as there are enough data points available. The RS predicts a maximum given by Eq. (2.16) which may indicate a region where to search for a local optimum by a gradient method. The efficiency of such statements depends on the quality of the approximation that is given by Eq. (2.18) or corresponding error estimators.

The RS must not always take into account all the allowable parameter combinations that may occur in the study. It might be better to limit the data used to some region where some or many promising solutions have been found.

Some care has to be taken to avoid clustering of the evaluation positions $\mathbf{P}_i$ in the parameter space. The parameters should cover the interesting region in a less or more homogenous way. On the other hand, a too regular grid of points may produce problems of pseudo-representations (cf. Sect. 2.8). Therefore, anti-clustering schemes like the Latin Hyper Cube are often used to produce strong but relatively small sets of data points.

Example 2.4 To provide a more realistic problem, we try to accelerate the optimization of the can extrusion (cf. Sect. 5.2.2). We have nine optimization parameters, so the number of coefficients of the RS (cf. Eq. 2.18) is

$$n_{coeff} = 1 + 9 + \frac{9(9+1)}{2} = 55.$$

We should use about

$$ndata \approx 2 * n_{coeff} \approx 100$$

variants.

These 100 input data should suffice to have a nice representation of the data by the RS. In consequence we do the first 100 studies, e.g. using a PSO optimization scheme. From these 100 jobs, we build the RS and find the optimum by Eq. (2.16). Continuing the search, we produce more results or variants of the design by the

Fig. 2.25 Apply meta model improvement to can backwards extrusion

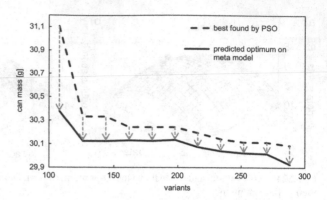

PSO. After some steps, we take the new results, replace the worst from the initial set and perform another RS-study. This procedure is repeated for some steps. Figure 2.25 indicates, that the meta model predicts designs that are essentially better than the ones found by the PSO. So we either come to faster predictions of good designs or find interesting improvements that are superior to the PSO-results after long searches.

In a broader sense we could think of an artificial neural net (ANN, cf. Sect. 2.4) as a meta model too. After some training, the ANN produces function values at given input of the free variables. The ANN are not of a low-level polynomial type but can approximate any function that depends on the free parameters. The quality of prediction of the functions value depends on the coverage of the parameter range and is measured during the testing phase.

2.8   Random or Deterministic Methods

Rolf Steinbuch

In Bionic Optimization we are in the field of non-deterministic methods, where the next step in a sequence is not necessarily known in the preceding step. Therefore, we should pay some attention to indeterminate problems. Certainly, we are not able to discuss all the problems related to stochastic processes and the specific ways to handle them. Other authors (e.g. Doltsinis 2012; Dreo and Siarry 2007) discuss these aspects in greater detail. Here we focus on some basic phenomena that are known to cause trouble with some optimizations.

All Bionic Optimization is at its core a random-based strategy. Variants of initial designs or drafts are created by some random modification of their parameters. Even if we use many different methods to avoid the search of the whole range of allowable solutions, we do not know where our next versions will be placed in the parameter space. Therefore, some discussion of random-based methods might help to understand the difference between deterministic methods such as the gradient based optimization and the bionic ones.

All random-based methods are hindered by slow convergence. From basic statistics, we remember that there are laws that relate errors to the number of tests:

$$error \sim \frac{1}{\sqrt{n}},$$ (2.19)

where n is the number of tests to be taken into account. So to reduce the error by 50 %, we need to increase the number of tests by a factor of 4. Furthermore, according to Eq. (2.19), we state that the ratio between error and number of trials for a given problem is roughly constant.

$$c \approx error_i * \sqrt{n_i}$$ (2.20)

When using random-based methods, people tend to introduce accelerating strategies to reduce the number of tests or, in our case of function evaluations, of time-consuming numerical simulations. The unthinking search of blind stochastic processes often takes too many trials to find good results. Intelligent users under pressure of time and budgets hope to replace it by faster, more intelligence-based methods.

Example 2.5 The Monte-Carlo integration of

$$I_{ex} = \int_0^1 e^x dx = e^1 - 1 = 1.71827\ldots$$

may help to understand the problem. Monte-Carlo-Integration is done by placing n_t random points in the rectangle $[0, 1] \times [0, 3]$ (Fig. 2.26a). We take the relative part n_b of points below e^x as measure of I_{ex}.

$$I_{ex} \approx 3\frac{n_b}{n_t}$$

From Table 2.5 we learn that the error follows Eq. (2.19), so we need to increase the number of function evaluations by a factor of 4 to reduce the error of the estimate by a factor of 2 as we expected from our knowledge of statistics.

If we place a regular grid of $n_x \times n_y$ points over the rectangle $[0, 1] \times [0, 3]$ (Fig. 2.26b) and quadruple the number of points from step to step, we realize that the error decreases proportionally close to the number of steps. We can guess more efficiently by defining a linear or parabolic shaped segment around the values of e^x and sub-sectioning there. The dashed parallelogram in Fig. 2.26b indicates such a region. It is evident, that, depending on the shape of the segment, the error would

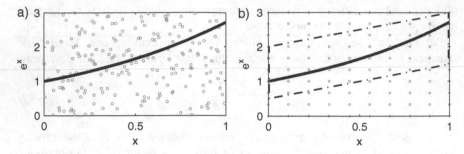

Fig. 2.26 Random and regularly spaced integration points. (**a**) Random points to guess integral. (**b**) Regularly spaced grid for integration, including a local sector to accelerate the convergence

Table 2.5 Averaged absolute	n	$error$	$\sqrt{n}$	$c \approx error^*\sqrt{n}$
error (100 runs) for different	100	0.1254	10.00	1.25
number of trials	200	0.0874	14.14	1.24
	400	0.0584	20.00	1.17
	800	0.0401	28.28	1.13
	1600	0.0315	40.00	1.25
	3200	0.0206	56.57	1.17

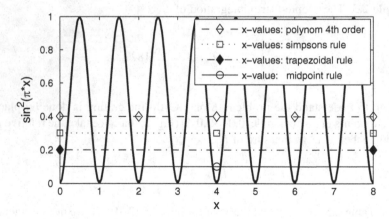

Fig. 2.27 Plot of $\sin^2(\pi x)$

decrease even faster. The consequence is that we often tend to avoid purely random methods. Unfortunately, such regular and intelligent approaches may lead to totally erratic results. A simple example may help us to understand the danger of such intelligent approaches.

Example 2.6 Let's find the integral of a set of data which we may not access easily but follows the (to us unknown) law (Fig. 2.27)

$$y = \sin^2(\pi x), \, x = 0 \ldots 8.$$

There is no doubt that

$$I_{ex} = \int_0^8 \sin^2(\pi x)dx = 4$$

is the exact value of the integral. Doing a numerical study, we start by taking the value of $x = 4$, the centre of the interval as the starting value, multiply it by the length of the interval and find $I_0 = 0$ (midpoint rule or rectangle method). Using the trapezoidal rule yields

$$I_1 = \frac{1}{2}\left(\sin^2(0) + \sin^2(8\pi)\right) * 8 = 0. \tag{2.21}$$

Then, using Simpson's rule yields

$$I_2 = \frac{1}{6}\left(\sin^2(0) + 4 * \sin^2(4\pi) + \sin^2(8\pi)\right) * 8 = 0. \tag{2.22}$$

The x-values of the numerical integrations are indicated in Fig. 2.27; they all are positioned at roots of $\sin^2(\pi x)$. Further increasing the polynomial order from 2 to 4 or subdivision of the interval yields $I_{num} = 0$ (Table 2.6) for at least two more steps, so we might accept it. So our intelligent and correct results are all wrong.

If we take a set of random x-values in the interval $[0, 8]$ and multiply the average function value by the length of the interval to estimate the value of the integral, we learn from Table 2.7 that there is a quick realistic guess of the true value. But even if

Table 2.6 Misleading numerical integration of a simple function

Degree of integration	Label of method	Value of integral
0	I_0	0
1	I_1	0
2	I_2	0
4	I_3	0

Table 2.7 Function integration using random x-values

# of x-values	Value of integral	Error (%)
1	1.742	56.445
2	0.904	77.403
4	3.530	11.750
8	3.999	0.020
16	4.237	5.920
32	4.434	10.843
64	4.076	1.903

the example shows nice data, the convergence to more exact values still follows the error estimator from Eq. (2.19).

To avoid any confusion, it must be mentioned that really intelligent guesses would use the benefit of the periodic characteristic of the sine function.

$$\int_0^8 \sin^2(\pi x)dx = 16\int_0^{1/2} \sin^2(\pi x)dx = 4. \tag{2.23}$$

As a consequence of this experience, it should be mentioned that deterministic procedures generally are faster than stochastic ones. But their tendency to be based on regular divisions inhibits the danger of failure, if the studied data follows a corresponding regularity. A general rule could be to use either deterministic or stochastic methods, but to be very careful when mixing them.

Example 2.7 Let's return to the response surfaces (RS) to discuss a more efficient way to solve the integral. We want to show how to use RS to approximate e^x by a small number of data points and use the approximation for an estimate of

$$I_{ex} = \int_0^1 e^x dx = e^1 - 1 = 1.71827\ldots. \tag{2.24}$$

In the range of $(x, y) = [0, 1] \times [0, 3]$, we define a set of *nrand* data points by a random definition or by a more qualified scheme such as Latin Hypercubes (McKay et al. 1979). For each of the points (x_i, y_i), we define a value

$$z_i = y_i - e^{x_i}. \tag{2.25}$$

These triples (x_i, y_i, z_i) are used to define a polynomial in x and y using a least square method or another appropriate scheme:

$$z(x, y) = a_0 + a_1x + a_2y + a_3x^2 + a_4xy + a_5y^2 \tag{2.26}$$

As we have six coefficients $a_0 \ldots a_5$ to be determined, *ndata* = 12 (at least twice the number of coefficients) would be a sufficient number of data points. From $z(x, y)$ = 0 in Eq. (2.26), we find a curve $y(x)$ that approximates e^x. Using some data on this curve allows for an approximation of the value of the integral. For 11–21 points and using Simpson's integration for the data, the error is in the range of 2–5 %. The use of RS is essentially more efficient than the random methods above, at least in this straightforward case. Figure 2.28 illustrates the process.

This is not very surprising, as seen in the Example 2.5, where we take into account only whether a point lies below or above the e^x-function. Here, we use the distance from the function to improve our guess. Such an approach corresponds to the increasing of the integration scheme in numerical integration, e.g. using Simpson's rule instead of the trapezoidal rule.

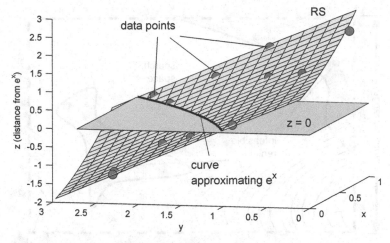

Fig. 2.28 Approximation of e^x by a response surface to integrate the function

For our optimization problems we conclude that we always consider whether it is appropriate to use bionic, random-based methods. If we do so, we employ them only up to a certain point. If we realize that our method approaches a local optimum, it could be a good idea to switch to a deterministic method to locate the exact value of the optimum. But when we continue searching other optima, we return to the bionic method we feel is adequate to the problem.

2.9   Violation of Boundary Conditions

Simon Gekeler

Beside keeping our optimization process to explore the given search space, defined by the parameters' upper and lower bounds, usually we need to comply with further constraints to get an acceptable and feasible optimization result. In all sequences of parameter sets based on random input, there is the potential of violating restrictions or boundary conditions. There are problems related to the fact that some parameter combinations cause infeasible geometries or unwanted geometry shapes, such as too small wall thicknesses. Exceeding limits on physical responses, such as the maximum stress or displacement, have to be taken into account as well. Thus, as indicated in Fig. 2.29, our search space may contain forbidden regions and we do not know where they are located.

The compliance of the defined parameter limits is not an onerous effort as we can directly influence the generation of new variants or check the proposals of the optimization algorithm before there is the computation of the new variants. Parameter values outside of the defined range can be corrected to keep the individuals

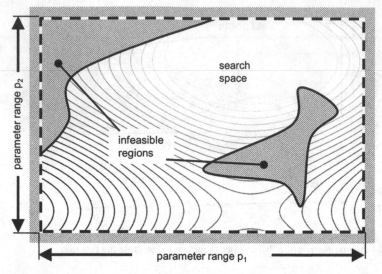

Fig. 2.29 Search space with infeasible combinations of the parameter values

exploring the desired region. The easiest way is to reset a lost individual exactly on
the exceeded parameter boundary. When doing this intervention, furthermore, the
characteristic of each optimization procedure should be considered. For the PSO
algorithm, for example, we recommend the modification of the particles' current
velocity vector as well, as otherwise the inertia value drives the particle in the next
iteration above the parameter limit again, and this decreases the performance of the
optimization process (cf. Eq. 2.1).

We can only identify unacceptable parameter sets in the search space after the
computation of new designs. To recognize violations in the optimization process, in
general, every restriction, such as stress or displacement limits, is defined with an
inequality constraint represented by the formulation

$$g_j(\mathbf{p}) \leq 0$$

with

$$j = 1 \ldots m,$$

where g represents the restriction value depending on the optimization parameters
$\mathbf{p}$ and m is the number of restrictions for one optimization problem. To deal with
violated inequality constraints there are many different methods available (Koziel
and Michalewicz 1999; Gekeler et al. 2014).

The easiest way is to *remove all unacceptable individuals* from the population
and to continue to produce members of the respective set until the required number
of acceptable individuals is found. There is no reason not to use this selection type

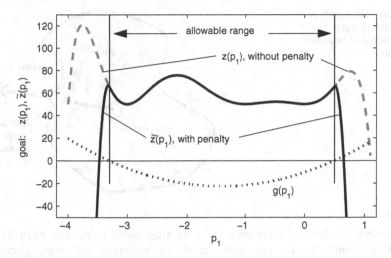

Fig. 2.30 Handling the violation of restriction $g(p_1) \leq 0$ with the penalty method when doing a maximization problem

unless the cost of a specific function evaluation is too extensive to produce a number of individuals that may be essentially larger than the number of usable individuals for the optimization process. Typical examples for such expensive individuals are non-linear FE-studies, where it takes some hours of computing time to find each specific solution.

Another way to keep the population near to the feasible range is to *punish all violations* of the given restrictions (Fig. 2.30). A penalty value *pen* weights the intensity of violation. This penalty is added to or subtracted from the objective z of the individual (Schumacher 2013). Then the optimization problem is

$$max\, \tilde{z}\,(\mathbf{p}) = z(\mathbf{p}) - r * pen(\mathbf{p})$$

and use, e.g.

$$pen(\mathbf{p}) = \sum_{j=1}^{m} max\Big(g_j(\mathbf{p}),0\Big)^2, \qquad (2.27)$$

where r is an additional user defined weighting. In consequence the individual is less attractive for the optimization procedure, while the non-punished individuals have better chances to reproduce in EVO or act as an attractor in PSO. On the other hand, punished individuals can also temporarily lead others to promising regions. Thus every computed design offers information about the search space. Furthermore, the penalty method allows for the approach to an optimum, located near a restriction boundary, from the infeasible region, too. The main disadvantage is that an optimization method with the standard penalty method does not guarantee

Fig. 2.31 Limitation of the individuals jump to handle the violation of boundary conditions

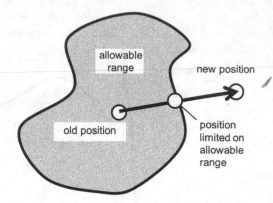

feasible results. But by increasing the weighting factor r, we can intensify the punishment and raise the probability to obtain acceptable outcomes. There are many variants and specific modifications in literature that may improve the performance of the various optimization procedures, e.g. the multiple linear segment penalty function for PSO, proposed by Plevris and Papadrakakis (2011).

A third idea among many others is to fix the parameters of the violating individual on the border of the allowable space (Fig. 2.31). This may be easily done for geometric input, but can be difficult if derived values such as stresses or displacement have to be considered. In such cases, reducing the mutation from the good parents' data in EVO or limiting the particles' velocity in PSO may be used. If the parameters change less, then the objective and the derived values will change less as well, so the violation may be avoided.

References

Au, S. K., & Beck, J. L. (1999). A new adaptive importance sampling scheme for reliability calculations. *Structural Safety, 21*, 135–158.

Bendsoe, M., & Sigmund, O. (2003). *Topology optimization: Theory, methods and applications.* Berlin: Springer.

Bourinet, J.-M., Deheeger, F., & Lemaire, M. (2011). Assessing small failure probabilities by combined subset simulation and Support Vector Machines. *Structural Safety, 33*, 343–353.

Chelouah, R., & Siarry, P. (2000). A continuous genetic algorithm designed for the global optimization of multimodal functions. *Journal of Heuristics, 6*, 191–213.

Das, P. K., & Zheng, Y. (2000). Cumulative formation of response surface and its use in reliability analysis. *Probabilistic Engineering Mechanics, 15*, 309–315.

Doltsinis, I. S. (2012). *Stochastic methods in engineering.* Southampton: WIT Press.

Dorigo, M., de Oca, M. A. M., & Engelbrecht, A. (2008). Particle swarm optimization. *Scholarpedia, 3*, 1486.

Dorigo, M., & Stützle, T. (2004). *Ant colony optimization.* Cambridge, MA: MIT Press.

Dreo, J., & Siarry, P. (2007). Stochastic Metaheuristics as sampling techniques using swarm intelligence. In T. S. C. Felix & K. T. Manoj (Eds.), *Swarm intelligence, focus on ant and particle swarm optimization*. Vienna, Austria: InTech Education and Publishing. doi:10.5772/ 5105. Available from: http://www.intechopen.com/books/swarm_intelligence_focus_on_ant_ and_particle_swarm_optimization/stochastic_metaheuristics_as_sampling_techniques_using_ swarm_intelligence. ISBN 978-3-902613-09-7.

Dubourg, V., Sudret, B., & Deheeger, F. (2013). Metamodel-based importance sampling for structural reliability analysis. *Probabilistic Engineering Mechanics, 33*, 47–57.

Fahlman, S., & Lebiere, C. (1991). *The cascade-correlation learning architecture*. Carnegie-Mellon University Pittsburg School of Computer Science.

Femur. *Femur fibre arrangement for strength, picture from: Wikimedia Commons (26.05.15)*. Original: Popular Science Monthly Volume 42.

Gekeler, S. (2012). *Die Partikelschwarmoptimierung in der Strukturmechanik*. Master thesis, Reutlingen University.

Gekeler, S., Pandtle, J., Steinbuch, R., & Widmann, C. (2014). Remarks on the efficiency of bionic optimisation strategies. *Journal of Mathematics and System Science, 4*, 139–154.

Gen, M., & Cheng, R. (2000). *Genetic algorithms and engineering optimization*. New York: Wiley.

Glover, F., & Laguna, M. (1999). *Handbook of combinatorial optimization* (pp. 2093–2229). New York: Springer US.

Günther, D., & Wender, K. F. (2010). *Neuronale netze*. Würzburg: Huber Hans.

Haftka, R. T., & Gürdal, Z. (1992). *Elements of structural optimization*. Dordrecht: Kluwer.

Hagan, M. T., & Menhaj, M. B. (1994). Training feedforward networks with the Marquardt algorithm. *IEEE Transaction on Neural Networks, 5*, 989–993.

Hansen, P., & Mladenović, N. (2005). *Variable neighborhood search* (pp. 211–238). New York: Springer US.

Hebb, D. O. (1949). *The organization of behavior. Bulletin of mathematical biophysics* (Vol. 5). New York: Wiley.

Kennedy, J. (1999). Small worlds and mega-minds: Effects of neighborhood topology on particle swarm performance. *Proceedings of the 1999 Congress on Evolutionary Computation, 3*, 1931–1938.

Kennedy, J. (2003). Bare bones particle swarms. *Proceedings of the IEEE Swarm Intelligence Symposium*. S. 80–87.

Kennedy, J., & Eberhardt, R. (1995). Particle swarm Optimization. *IEEE International Conference on Neural Networks* (pp. 1942–1948).

Kennedy, J., & Eberhardt, R. (1997). A discrete binary version of the particle swarm algorithm. *Proceedings of the IEEE International Conference on Systems, Man and Cybernetics* (pp. 4104–4108).

Koziel, S., & Michalewicz, Z. (1999). Evolutionary algorithms homomorphous mappings, and constrained parameter optimization. *MIT Press Journals, 7*, 19–44.

Kramer, O. (2009). *Computational intelligence: Eine Einführung*. Berlin: Springer.

Lagaros, N. D., & Papadrakakis, M. (2004). Learning improvement of neural networks used in structural optimization. *Advances in Engineering Software, 35*, 9–25.

Matthies, H. G., Litvinenko, A., Rosić, B. V., Kučerová, A., Sýkora, J., & Pajonk, O. (2013). Stochastic setting for inverse identification problems. *Report to workshop numerical methods for PDE constrained optimization with uncertain data*. Report No. 04/2013. DOI:10.4171/ OWR/2013/04, 27 January 2013

McKay, M. D., Beckman, R. J., & Conover, W. J. (1979). A comparison of three methods for selecting values of input variables in the analysis of output from a computer code. *Technometrics, 21*, 239–245.

Mendes, R., Kennedy, J., & Neves, J. (2004). The fully informed particle swarm: Simpler, maybe better. *IEEE Transactions on Evolutionary Computation, 8*, 204–210.

Perez, R. E., & Behdinan, K. (2007). Particle swarm optimization in structural design. In F. T. S. Chan & M. K. Tiwari (Eds.), *Swarm intelligence: Focus on ant and particle swarm optimization*.

Plevris, V., & Papadrakakis, M. (2011). A hybrid particle swarm—Gradient algorithm for global structural optimization. *Computer-Aided Civil and Infrastructure Engineering, 26*, 48–68.

Rechenberg, I. (1994). *Evolutionsstrategie '94*. Stuttgart: Frommann-Holzboog.

Rozvany, G. I. N. (1997). *Topology optimization in structural mechanics*. Vienna: Springer.

Schumacher, A. (2013). *Optimierung mechanischer Strukturen*. Berlin: Springer.

Schwefel, H. P. (1989). *Numerical optimization of computer models*. Chichester: Wiley.

Shi, Y., & Eberhart, R. (1998). A modified particle swarm optimizer. IEEE World Congress on Computational Intelligence. *Evolutionary Computation Proceedings* (pp. 69–73).

Smith, A. R., Pryer, K. M., Schuettpelz, E., Korall, P., Schneider, H., & Wolf, P. G. (2006). A classification for extant ferns. *Taxon, 55*, 705–731.

Steinbuch, R. (2010). Successful application of evolutionary algorithms in engineering design. *Journal of Bionic Engineering, 7*(Suppl), 199–211.

Stock, P., & Zülch, G. (2012). Reactive manufacturing control using the ant colony approach. *International Journal of Production Research, 50*, 6150–6161.

Vanderplaats, G. N. (1984). *Numerical optimization techniques for engineering design*. New York: McGraw-Hill Ryerson.

Widrow, B., & Hoff, M. E. (1960). Adaptiv switching circuits. IRE WESCON Convention Record.

Zell, A. (1997). *Simulation neuronaler Netze*. München: Oldenbourg.

Chapter 3
Problems and Limitations of Bionic Optimization

Tatiana Popova, Iryna Kmitina, Rolf Steinbuch, and Simon Gekeler

We have seen that Bionic Optimization can be a powerful tool when applied to problems with non-trivial landscapes of goals and restrictions. This, in turn, led us to a discussion of useful methodologies for applying this optimization to real problems. On the other hand, it must be stated that each optimization is a time consuming process as soon as the problem expands beyond a small number of free parameters related to simple parabolic responses. Bionic Optimization is not a quick approach to solving complex questions within short times. In some cases it has the potential to fail entirely, either by sticking to local maxima or by random exploration of the parameter space without finding any promising solutions. The following sections present some remarks on the efficiency and limitations users must be aware of. They aim to increase the knowledge base of using and encountering Bionic Optimization. But they should not discourage potential users from this promising field of powerful strategies to find good or even the best possible designs.

3.1 Efficiency of Bionic Optimization Procedures

Iryna Kmitina and Tatiana Popova

Bionic Optimization strategies have proven to be efficient in many applications, especially where there are many local maxima to be expected in parameter spaces of higher dimensions. In structural mechanics, the central question is whether one particular procedure is to be preferred generally or if there are different problem types where some procedures are more efficient than others. Evolutionary

T. Popova • I. Kmitina • R. Steinbuch • S. Gekeler (✉)
Hochschule Reutlingen, Reutlingen Research Institute, Alteburgstraße 150, 72762 Reutlingen, Germany
e-mail: Tatiana.Popova@Reutlingen-University.DE; Iryna.Kmitina@Reutlingen-University.DE; Rolf.Steinbuch@Reutlingen-University.DE; Simon.Gekeler@Reutlingen-University.DE

© Springer-Verlag Berlin Heidelberg 2016
R. Steinbuch, S. Gekeler (eds.), *Bionic Optimization in Structural Design*,
DOI 10.1007/978-3-662-46596-7_3

Optimization with some sub-strategies, Particle Swarm Optimization, Artificial Neural Nets, along with hybrid approaches that couple the aforementioned methods have been investigated to some extent. These approaches are not uniquely defined, but rather imply many variants in the definition and selection of next-generation members, varying settings of the underlying processes, and criteria for changing strategies. Some simple test problems were used to quantify the performance of these different approaches. The measure of the procedures performance was the number of individuals which needed to be studied in order to come up with a satisfactory solution. As our main concern is problems with many parameters to be optimized, artificial neural nets do not show sufficient convergence velocities in our class of optimization studies to be included. Evolutionary Optimization, including its subclasses of Fern Optimization, and Particle Swarm Optimization prove to be of comparable power when applied to the test problems. It is important to note that, for all these approaches, some experience of the optimization parameters has to be gathered. In consequence, the total number of runs or individuals necessary to do the final optimization is essentially larger than the number of runs during this final optimization. Good initial proposals prove to be the most important factor for all optimization processes.

3.1.1 Comparing Bionic Optimization Strategies

As discussed in many sections of this book, Bionic Optimization may be defined by many different approaches. In this section, we deal with some of the most commonly accepted classifications, without taking into account all the many sub-classifications that might be found in the literature. The central approaches we compare are:

- Evolutionary Strategy (EVO, Sect. 2.1) (Rechenberg 1994; Steinbuch 2010)— where paired or crossed parents have children by the combination and mutation of their properties. These children, or some of them, are parents in the next generation.
- Fern Strategy (FS, Sect. 2.2)—which may be regarded as a simplification of Evolutionary Optimization. Individuals have offspring by mutation only, not by crossing properties with other members of the parent generation.
- Particle Swarm Optimization (PSO, Sect. 2.3) (Coelho and Mariani 2006; Plevris and Papadrakakis 2011)—where a population drifts through the possible solution space. The swarm's coherence is given by simple rules about the velocity of the individuals.
- Artificial Neural Nets (ANN, Sect. 2.4) (Berke et al. 1993; Lagaros and Papadrakakis 2004; Widmann 2012)—where training of the net yields an understanding of the solution space and allows the prediction of the system's response to given input. As ANN are not very efficient when applied to problems with many free parameters, we do not discuss them here (Widmann 2012).

3.1.2 Measuring the Efficiency of Procedures

To quantify the efficiency of the different optimization strategies, we must introduce a measure that allows us to uniquely define the amount of work required to achieve a predefined quality. From previous experience, we propose using the number of individuals to be analyzed before coming close to an accepted good value. This requires knowledge of what a good solution would be, which is generally not known when we start studying new problems.

To consider the violation of boundary conditions (cf. Sect. 2.9) we restrict our present study to the use of penalty functions. The geometric input is set to the minimum or maximum value, if the randomly produced data exceed the respective limits. For PSO, we invert the particle's previous velocity, if it violates given limits in addition to the penalty value. This combined approach has the advantage of simple applicability.

3.1.3 Comparing the Efficiency of Bionic Optimization
Strategies

Optimization is an expensive and time consuming process. We need to understand which procedure and which combinations of parameters may lead to a good and acceptable result within a reasonable amount of time.

Test Examples

Figure 3.1 depicts the five test examples used while Table 3.1 summarizes their data. We want to minimize the mass of the frames by varying the rods' cross sections without exceeding their maximum stresses and displacements. The grid size of the examples is 1000 mm, except for example F2 where the grid size is 360 in. Example F2 used imperial units (in., kip) the other frames use mm and Newton.

To come up with comparable results, we performed a series of 20 loops for each problem and each strategy to avoid having only one or few very good or very bad results. Also, the optimization settings we used were based on previous experience with the underlying problems, so the number of runs presented does not come from naïvely starting a procedure, but includes some preliminary work which is impossible to quantify.

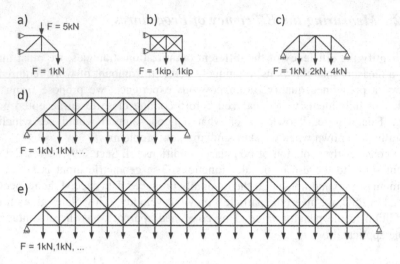

Fig. 3.1 Test frames with loads and supports. (**a**) F1: 6 rods frame. (**b**) F2: 10 rods frame. (**c**) F3: 13 rods frame. (**d**) F4: 58 rods frame. (**e**) F5: 193 rods frame

Table 3.1 Data of test problems

Frame	Free param.	Grid size	Amax/Amin	E-Mod	σ_{max}	d_{max}
F1	6	1000 mm	600/20 mm^2	200 GPa	120 MPa	0.5 mm
F2	10	360 in.	35/0.1 in.^{2}	10 Msi	25 ksi	2.0 in.
F3	13	1000 mm	400/20 mm^2	200 GPa	50 MPa	0.5 mm
F4	58	1000 mm	400/20 mm^2	200 GPa	100 MPa	2.0 mm
F5	193	1000 mm	600/20 mm^2	200 GPa	450 MPa	20 mm

Parameters: # of rods in frame; grid size: horizontal or vertical distance between the nodes; Amax, Amin: maximum and minimum allowed cross section area of the rods; E-Mod: Young's modulus; σ_{max}: maximum allowed stress in rod; d_{max}: maximum allowed displacement of nodes

Input and Results of the Test Examples

Tables 3.2, 3.3 and 3.4 list the inputs of the test runs used. Table 3.5 and Fig. 3.2 (individuals per loop) summarize the results of the test runs. The most important data are the number of individuals analyzed to find a sufficient good design labelled as 'Individuals [1000]'. The number given multiplied by 1000 gives the total number of individuals required to find the proposed design. *mean* and *stddev* (standard deviation) and *best* are descriptions of the results of the 20 runs. The ratio of the difference between the best and the average result divided by the standard-deviation (*reldev*) gives an idea of the stability of the strategy.

Table 3.2 Optimization settings used for EVO

Model	Parents	Kids	Mut. rad. max	Mut. rad. min	Generations
F1	10	20	0.5	0.05	60
F2	5	10	0.5	0.05	40
F3	5	10	0.5	0.05	50
F4	50	100	0.5	0.05	100
F5	100	200	0.5	0.05	200

Table 3.3 Optimization settings used for FS

Model	Parents	Kids/parent	Mut. rad. max	Mut. rad. min	Generations
F1	10	5	0.5	0.05	100
F2	10	4	0.5	0.05	50
F3	20	5	0.5	0.05	100
F4	100	5	0.5	0.05	200
F5	200	5	0.5	0.05	200

Mutation radius reduced for EVO and FS: 0–25 % of generations: $r_{mut} = 0.50$, 25–50 % of generations: $r_{mut} = 0.20$, 50–75 % of generations: $r_{mut} = 0.10$, 75–100 % of generations: $r_{mut} = 0.05$

Table 3.4 Optimization settings used for PSO

Model	Particles	Generations
F1	10	100
F2	10	50
F3	20	100
F4	100	200
F5	200	200

Weighting factors: $c_1 = 0.08$, $c_2 = 0.005$, $c_3 = 2.0$

Interpretation of the Results

EVO, FS and PSO prove to be of a comparable efficiency when applied to the four smaller problems (F1, F2, F3, F4). Figure 3.2 indicates that there might be a nearly linear relation between the number of optimization variables and the individuals required to find good proposals. For the largest problem, F5, FS displays a performance that is essentially weaker than EVO and PSO. EVO and PSO seem to be of comparable power when applied to the problem class which we discuss here. FS shows promising results if the number of free parameters is not too large, but is less successful in random search in high dimensional spaces. The scatter indicator *reldev* proposes that PSO has a more stable tendency to find solutions near the best, while EVO and FS show a larger range after the 20 runs.

Some knowledge may be gleaned from the results of these series of studies. Foremost, that optimization, especially Bionic Optimization, is a process that consumes large amounts of time and computing power. Furthermore the results

Table 3.5 Results of 20 optimization runs per problem

Strategy	Model	Mean	Stddev	Best	Reldev	Individuals [1000]
EVO	F1	1.62e6	0.716e3	1.62e6	1.50	12
	F2	6.33e4	4.50e3	5.47e4	1.90	8
	F3	2.56e6	6.99e4	2.48e6	1.11	20
	F4	1.03e7	4.18e5	8.65e6	3.92	200
	F5	1.98e7	1.07e6	1.58e7	3.65	800
FS	F1	1.66e6	4.49e4	1.62e6	0.81	28
	F2	6.39e4	4.43e3	5.45e4	2.09	25
	F3	2.50e6	2.29e4	2.47e6	1.19	46
	F4	9.91e6	2.77e5	9.39e6	1.86	189
	F5	2.33e7	4.18e5	2.25e7	2.15	2570
PSO	F1	1.65e6	1.71e4	1.62e6	1.61	6
	F2	5.87e4	5.61e3	5.15e4	1.27	8
	F3	2.50e6	2.53e4	2.48e6	1.02	32
	F4	8.90e6	1.68e5	8.68e6	1.22	200
	F5	1.54e7	0.18e4	1.53e7	1.70	1120

Fig. 3.2 Efficiency of the three Bionic Optimization strategies tested

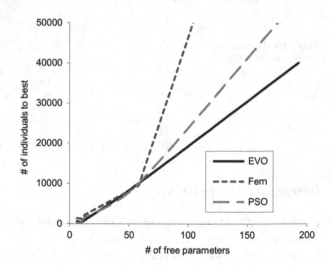

presented in this section would not have been found without a large number of preliminary studies providing experience in the field of frame optimization.

The input characteristics used in the test runs is derived from these preliminary studies. For example the selection of the three weighting factors $\{c_1, c_2, c_3\}$ for the PSO required 100,000 runs. The proposal of the reduction of the mutation range for EVO and FS is the result of many studies as well. The proposal to use a number of initial parents or individuals in the size of the number of free variables for EVO and PSO is based on many studies, as well as the idea to use a large number of initial parents and a small number of children in FS.

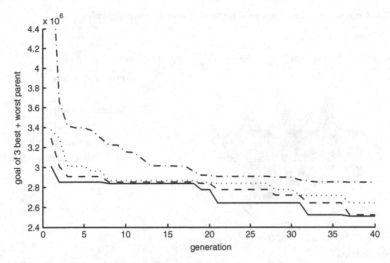

Fig. 3.3 History of an Evolutionary Optimization for example F3 (Fig. 3.1)

One central fact about all optimization may be learned from Fig. 3.3. If there is a good initial design, the number of optimization runs to be done may decrease significantly. If an experienced engineer proposes an initial design with a goal of e.g. 2.7×10^6, we need only 20 generations or 50 % of the workload required to solve the task with a random initial design.

Further Test Examples

In addition to the trusses (Fig. 3.1) representing static structural optimization problems, there are many other test examples available (Surjanovic and Bingham 2015). Most of them are defined by mathematic functions and because of their characteristics and the known data of global and local optima, they are most suitable for algorithm testing. A selection of such benchmark problems is listed in Table 3.6. Some of them can be expanded to an arbitrary number of dimensions d. Especially in field of optimization they are useful in algorithm development or to improve existing procedures. Furthermore, with awareness of these problems, users can gain experience in optimization strategies, check their efficiency and learn how to choose proper optimization settings.

3.1.4 Conclusions

The quality of the initial proposals is the most important component of any optimization. If experienced and motivated engineers propose designs that are close to the optimal ones, there is a good chance that at least a local optimum

Table 3.6 Common test functions used for testing optimization algorithms

Eggholder function

$$f(p_1, p_2) = -(p_2 + 47)sin\left(\sqrt{\left|p_2 + \frac{p_1}{2} + 47\right|}\right) - p_1 sin\left(\sqrt{|p_1 - (p_2 + 47)|}\right)$$

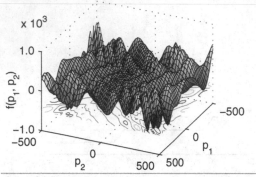

Free parameters:
$i = 1, 2$
Search domain:
$p_i \epsilon$ [−512, 512]
Global optimum (min):
$f(512, 404.2319) = -959.6407$

Schwefel function

$$f(\mathbf{p}) = 418.9829 * d - \sum_{i=1}^{d} p_i * sin\left(\sqrt{|p_i|}\right)$$

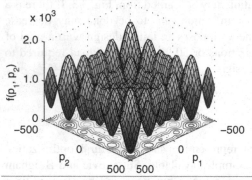

Free parameters:
$i = 1, \ldots, d$
Search domain:
$p_i \epsilon$ [−500, 500]
Global optimum (min):
$f(p_i = 420.9687) = 0$

Ackley function

$$f(\mathbf{p}) = -20\,exp\left(-0.2\sqrt{\frac{1}{d}\sum_{i=1}^{d} p_i^2}\right) - exp\left(\frac{1}{d}\sum_{i=1}^{d}\cos\left(2\pi p_i\right)\right) + 20 + exp(1)$$

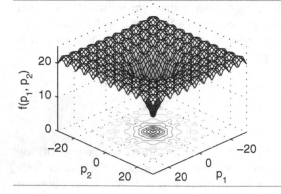

Free parameters:
$i = 1, \ldots, d$
Search domain:
$p_i \epsilon$ [−32.768, 32.768]
Global optimum (min):
$f(p_i = 0) = 0$

(continued)

Table 3.6 (continued)

Goldstein-Price function

$$f(p_1, p_2) = \left[1 + (p_1 + p_2 + 1)^2 \left(19 - 14p_1 + 3p_1^2 - 14p_2 + 6p_1 p_2 + 3p_2^2\right)\right]$$
$$* \left[30 + (2p_1 - 3p_2)^2 \left(18 - 31p_1 + 12p_1^2 + 48p_2 - 36p_1 p_2 + 27p_2^2\right)\right]$$

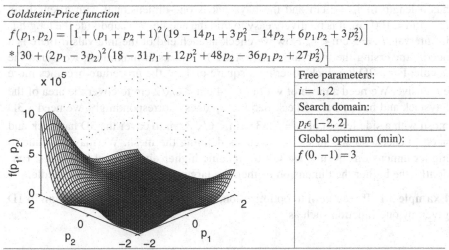

Free parameters:

$i = 1, 2$

Search domain:

$p_i \in [-2, 2]$

Global optimum (min):

$f(0, -1) = 3$

will be found which is not too far away from the best solution possible. If we are close to good proposals, gradient methods will improve the free parameters in a short time and with reasonable effort.

As soon as we doubt that our initial designs are close to the optimal ones, EVO or PSO have the capacity to propose better designs. Nevertheless, the number of function evaluations may be large. Which of the two is preferred for a particular problem must be decided with some preliminary test. Often, the particle swarm shows a faster tendency towards the assumed best values, but some examples indicate that the swarm might have the tendency to stick to local maxima, just as do gradient methods.

Switching to Gradient Optimization if approaching a maximum closely is always an interesting option. But experience of the problem and methodology is required there as well.

In every case, the optimization of large problems consumes time and resources. There is no way to avoid the evaluation of many individual solutions and there is no guarantee that the absolute best solution will be found at all.

3.2   The Curse of Dimensions

Rolf Steinbuch

One of the most problematic properties of optimization tasks with higher numbers of free parameters is "the curse of dimensions". It appears to be one of the most governing drawbacks and sets the strongest limits on any attempt to accelerate the progress of all optimization strategies when dealing with higher numbers of free parameters. We may compare it to the search for a needle in a haystack. If the needle has a length of $l_n = 6$ cm and the haystack is one-dimensional with a length on $l_h = 1$ m $= 100$ cm, it is relatively easy to find the needle. We subdivide the length l_h in intervals $l_{1i} = 5$ cm $< l_h$. Now we check at each end of the intervals if there is a needle traversing the interval border. After 19 checks at maximum, we found the needle. For a 2D haystack covering a square of 1 m² the procedure becomes more expensive. We need a mesh of width $l_{2i} = 4$ cm $< 6/\sqrt{2}$ cm to cover the area of the haystack and have now to check 625 mesh edges. Correspondingly, we need a 3D mesh with a side length of $l_{3i} = 3.333$ cm $< 6/\sqrt{3}$ cm to cover the 3D haystack and need to check 36,000 faces of the cubes defining the mesh. We may continue to higher dimensions even if we fail to imagine higher dimensional haystacks. Evidently, the higher the dimension is, then the larger the effort to find the needle.

Example 3.1 If we return to optimization, we may assume a local optimum in 1D given by one function such as

$$goal(p_1) = \frac{1}{2}(1 + \cos(\pi^* p_1))$$

visualized in Fig. 3.4a. In the interval $-1 < p_1 < 1$ the 2D area below the function is

$$V_1 = 1 = 0,5 * V_{0,1}, \tag{3.1}$$

where $V_{0,1} = 2$ the area of the surrounding rectangle. If we step to a 2D problem, the corresponding function becomes

$$goal(p_1, p_2) = \frac{1}{4}(1 + \cos(\pi^* p_1))(1 + \cos(\pi^* p_2)) \tag{3.2}$$

(Fig. 3.4b) and the volume below the hill is

$$V_2 = 1 = 0,25 * V_{0,2}, \tag{3.3}$$

where $V_{0,2} = 4$ the volume of the surrounding cube.

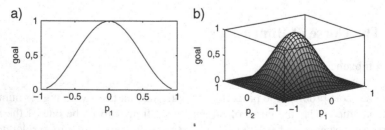

Fig. 3.4 Decreasing volume of hills covered by cos functions. (**a**) 1D cos function. (**b**) 2D cos function

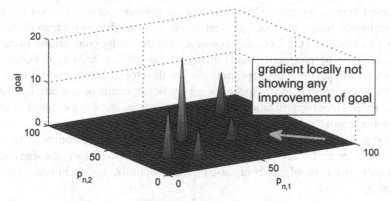

Fig. 3.5 Local optima as steep needles in higher dimensional spaces

We realize that the higher the number of dimensions n, the smaller the ratio

$$\frac{V_n}{V_{0,n}} = \left(\frac{1}{2}\right)^n. \tag{3.4}$$

The probability to find the hill gets smaller and smaller as the number of dimensions increases. The hills of local optima degenerate to needles, as Fig. 3.5 tries to demonstrate. Even worse is the fact that the gradient of the goal on the plane between the needles is close to zero; there is no strong force driving the optimization process in the direction of the needles. So we are bound to do many studies and repeat many searches to come close to promising designs. This problem is often called the curse of dimensions. It limits the maximum achievable velocity to first find a promising region and then converge to a solution. The only work-around is a reduction of the problem's dimension, which reduces the power of the optimization range or the reduction of the search space, which may exclude interesting regions. Therefore we have to live with the curse of dimensions if we are dealing with the optimization of problems with many free parameters.

3.3 Acceleration of Bionic Optimization Processes

Tatiana Popova

Optimization today is a very promising area and is used as a standard method to decrease production costs or the weight of a part or assembly. From the point of mechanical calculation, optimization methods enable designers to choose the best variant of a design with the best allocation of resources, reduction of the cost of materials, energy, and etc. All parts of the optimization procedure are important, from identification of variables and initial algorithm parameters to identification of the correct fitness function. Optimization as a subject receives serious attention from engineers, scientists, managers and anyone else involved with manufacturing, design, or business. This focus on optimization is driven by competition in quality assurance, cost of production and, finally, in success or failure of businesses. Throughout the past century, optimization has developed into a mature field that includes many specialized branches, such as linear conic optimization, convex optimization, global optimization, discrete optimization, etc. Each of these methods has a sound theoretical foundation and is supported by an extensive collection of sophisticated algorithms and software. With rapidly advancing computer technology, computers are becoming more powerful, and correspondingly, the size and the complexity of the problems being solved using optimization techniques are also increasing.

The requirements for optimization is the possibility of achieving good results within a short processing time. Gradient methods can be sufficient, but the increase of complexity of optimized components often leads gradient methods to wrong proposals at local optima. Gradient methods are sufficient when the task is to find a local optimum. Gradient methods are not applicable for global optima search. The methods stop at one hill of the goal function without investigating the others, which could possibly contain better results. This situation necessitates the investigation and research into new optimization methods that could deal with complicated optimization problems.

Traditional optimization algorithms often depend on the quality of the objective function, but many objective functions are usually highly non-linear, steep, multi-peak, non-differential or even discontinuous, and have many continuous or discrete parameters. Almost all problems need vast amounts of computation. Traditional optimization techniques are incapable of solving these problems. Bionic engineering copies living systems with the intention of applying their principles to the design of engineering systems. In recent years, bionic engineering has been actively developed globally. Much bionic scientific research has been conducted, and new products have been designed and developed. Biomimetic structural optimization methods, for example, aim at the improvement of design and evaluation of load-bearing structures.

The quality of algorithms depends not only on the problem's complexity, but also on the individual adjustment of the parameters of the corresponding optimization methods. Incorrect choices for algorithm parameters could lead to a decrease in search time or even to false results. This shows the importance of collecting the results of an algorithm's parameters variation to understand their influence. The investigation's goal is to find sets of parameters that could provide stable results for the wide problem range.

The efficiency of the algorithms should be proven. Optimization testing functions are used to estimate the quality of algorithms. These functions (e.g. in Table 3.6) were chosen because they represent the common difficulties seen in isolated optimization problems. By comparing and contrasting these functions, a user can make judgments about the strengths and weaknesses of particular algorithms.

Essential for optimization problems is the identification of the point when convergence is reached, so as not to lose time on future unnecessary investigations. This is especially valuable for optimizations with a high number of variables and relatively long fitness function evaluation time.

To find an acceptable optimization calculation time, accelerating strategies must be found. Some are based on the fact that not all parameters have the same impact on the object. Decreasing the number of parameters handled may help to reach areas with good objective values. Other strategies use statistical predictions to estimate the best values of the objective function in early stages. From those estimations, a decision on whether the reached objective value is promising could be derived. Unfortunately these predictions, again, need many runs to yield reliable data. Nevertheless, accelerated optimization may be a tool to find reliable results at acceptable time and cost.

3.3.1 Selecting Efficient Optimization Settings

Tatiana Popova

We have been discussing different ways to perform Bionic Optimizations. However, we are missing some guidelines for running a real task, i.e. how many parents, kids, individuals to use, how to define crossing and mutation, which weighting factors will perform well, and which will be less effective.

As there are unlimited possible problems and different strategies tend to perform differently in different applications, general rules are hard to propose. Nevertheless, here we try to give some hints for starting a process. Motivated users will soon learn how to accelerate their studies. They will optimize the optimization process, also known as meta-optimization. We discuss specific strategies for each of the three most important Bionic Optimization methods respectively.

Evolutionary Optimization

The optimization data we may access are:

- Number of parents
- Number of kids
- Survival of parents
- Way of crossing
- Mutation radius
- Number of generations

As a general rule, we realize, the larger the number of individuals, the greater the probability to find promising results. But the time required for analysis limits these numbers, so we need some starting data. From our previous experiences, we

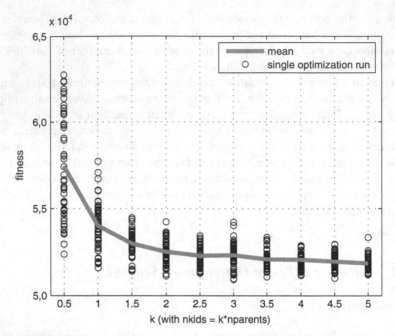

Fig. 3.6 Minimization of goal with multiple of number of kids related to # of parents. Taking # of kids $= 2 \times$ # of parents seems a good guess

propose to use a *number of parents* in the range of 0.5–2 times the number of free parameters. The higher the number of parents, the better the performance will be. From Fig. 3.6 we learn that the *number of kids* should be about double the number of parents, while less is not efficient, higher values seem also not to be as efficient because the computational effort increases with the rising number of individuals, without achieving much better results. In this study we used the 10 rods frame problem (Fig. 3.1b) and did multiple optimization runs with different number of kids, related to the number of parents, and fixed values for the other optimization settings (*nparents* $= 15$, *ngen* $= 40$, *mutrad* $= 25$ %, parents survive). For each variation, we repeated in a loop of 50 runs.

Survival of parents is generally recommended, as it removes the danger of deleting some good proposed designs. It might be a good idea to limit the number of generations an individual may survive, e.g., to a maximum of five generations, but this does not increase the performance of the process in a very powerful way.

Crossing might be accomplished by using the ideas outlined in Sect. 2.1. From our experience, averaging parents' properties is quite a good idea.

Mutation might be done to each parameter individually. Using large mutation radii leads to a pure random search, where the properties of the parents no longer influence those of the kids. Too small mutation radii correspond to a local search around the parents' values. If a local search is intended, it is better to use a gradient method. We generally recommend starting with a mutation radius in the range of 15–25 % of a parameter's range. After a certain number of generations, these mutation radii might be reduced to values of 5–10 %. A switch to a gradient method could be a good idea as well.

The *number of generations* should not be less than the number of free parameters. To avoid unnecessary repetitions, it is recommended to monitor the process and to kill it if no further progress is observed. We realize that the method does not often show any further improvement, so it is appropriate to stop the run. If restarting the process at any generation is possible, it might save time and computing power.

Fern Optimization

Fern Optimization is, from our experience, recommended for problems with small numbers of free parameters. Figure 3.2 using the models of frames introduced in Sect. 3.1 indicates that, for more than 50 parameters, PSO and EVO are essentially more efficient. Using Fern Optimization, we are concerned with:

- Numbers of parents
- Numbers of kids per parent
- Mutation radius
- Number of generations
- Killing underperforming families (cf. Fig. 2.6)

Again, the *number of parents* should not be essentially smaller than the number of free parameters, as this number supports the coverage of the parameter space.

The *number of kids* should be about 3–5 per parent; more is better, but increases computing time. These kids are generated by the mutation of one parent's properties. To reiterate, a not too large *mutation radius* of 15–25 % is recommended at the onset of the study. Keep in mind: as soon as good designs have been approached, the mutation radius can be decreased or the optimization could switch to a gradient method.

Additionally, it is important that a sufficient *number of generations* should be calculated. The process can be terminated if it approaches saturated levels, which can be monitored.

The decision to remove a family, the offspring of one initial parent from the process, is possible, but should not start too early as indicated by Fig. 2.6.

Particle Swarm Optimization

PSO generally tends to give very promising results as long as some basic criteria are met. The definition of the optimization parameters covers:

- Number of individuals
- Number of time steps
- Initial velocities
- Weighting of contributions to velocity update

The *number of individuals* seems to be the most important input for PSO. At least, the number of free parameters should be covered. Again, the larger the number, the better the results, but the larger the computational effort as well. On the other hand, we often observe the particle swarm sticking to local optima. This might be dependent on the number of individuals, so again, larger numbers are to be preferred. The *number of time steps* must not be too large, as often 10–20 generations yield interesting designs, and a restart option allows continuing processes that have not yet converged. The *initial velocities* should be in the range of 10–25 % of the parameter range for each of the optimization parameters.

Many discussions deal with the definition of the *weighting factors* in the equation (cf. Eq. (2.1) and the associated declaration)

$$\mathbf{v}_j(t+1) = c_1 \mathbf{v}_j(t) + c_2 \mathbf{r}_1 \circ (\mathbf{p}_{Pb,j} - \mathbf{p}_j(t)) + c_3 \mathbf{r}_2 \circ (\mathbf{p}_{Gb} - \mathbf{p}_j(t)).$$

Generally, a large value of c_1 (inertia) yields a broad search of the parameter space, but suppresses convergence to the best values. Large values of c_3 (social) accelerate the convergence, but might stick to early found local optima. The weighting of c_2 (cognitive) has a smaller effect on the performance. As a coarse rule we often use

- $0.4 < c_1 < 1$
- $0 < c_2 < 0.5$
- $0.5 < c_3 < 2$.

These proposals might help to get started more effectively using Bionic Optimization methods, but are not guaranteed to be the best ones for every problem.

3.3.2   Parallelization and Hardware Acceleration

Simon Gekeler

Using Bionic Optimization procedures requires computing and evaluating many different design variants. It is a time consuming process, but offers the chance to find the global optimum or at least a good solution, even in a large and complex design space with many local optima. If a single design evaluation, e.g., a non-linear FEM simulation, takes an excessive amount of computation time, it is no longer efficient to integrate a population-based optimization method, like PSO or EVO, into the usual design development process. The number of evaluated variants is limited by schedules and processing capabilities. We need strategies for applying these methods in an acceptable period of time.

To reduce process times, first check the model being optimized and its computation time, before trying to accelerate the optimization algorithm itself. In FEM simulations, for example, we should verify if the model is adequately simplified to be calculated quickly with sufficient result accuracy. Additionally, it is worthwhile to allot some time for an accurate parameterization of the design. Especially in case of geometry parameters, good parameterization is important for the prevention of inconsistent situations during automated variance of the parameters in the optimization process, which leads to aborted processes and needless waste of development time.

To perform a quick and efficient optimization run, which means to achieve satisfying results with fewer design computations, choosing the right strategy is critical, including the choice of optimization method and algorithm settings (see Sect. 3.1.1) for the specific type of problem. To reach that goal, there are different ways to proceed, e.g., run the optimization in several stages, maybe using different methods, limited parameter ranges, or just using the most significant parameters for an exploration phase first, and following with an detailed phase in a localized area but increased parameter space. Furthermore, hybrid optimization methods, such as PSO with an automated switch to the Gradient method in the final stage (Plevris and Papadrakakis 2011) or meta models (cf. Sect. 2.7), can greatly reduce optimization time. Preliminary investigations, such as sensitivity analysis, can help to reduce computation effort with more information about the problem, choosing the important parameters and neglecting the insignificant ones. For highly complex problems, where the computation time is expected to be large to find the global optimum, the most efficient method could be finding an already acceptable local optimum by using the relatively fast gradient method with a good starting position.

If the optimization job is well prepared and ready for efficient execution, we can possibly gain an additional and drastic reduction of processing time by accelerating the computation time with using faster computer hardware or with using further computer resources.

Parallel Jobs for Speeding Up Optimization Processes

In Bionic Optimization procedures, such as EVO or PSO, in one generation or iteration we compute many different design variants, before the design's results are evaluated and the next loop starts with the computation of newly generated designs Thus, in this section of the optimization algorithm, each job can be processed independently of each other. This allows for the possibility of running the jobs in parallel, enabling an enormous acceleration of these optimization procedures. The ability of optimization algorithms for parallelization depends on the ratio of the sequential workflow and tasks which can be done in parallel. For example, with EVO we can increase the number of parallel tasks by increasing the number of kids to be calculated in one generation. However, we rely on adequate optimization settings to improve calculation times.

When processing Bionic Optimization tasks in parallel, we are referring to parallel computation of design variants on different workstations connected by a Local Area Network (LAN). An architecture for the distribution of jobs to be run in parallel is depicted in Fig. 3.7. The optimization algorithm is running on one workstation (master), which distributes the tasks to be run in parallel to different

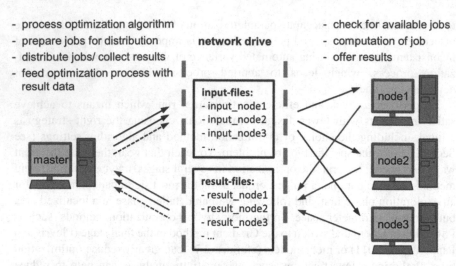

Fig. 3.7 Parallelization of optimization processes in a computer cluster

workstations (nodes) in the cluster. Afterward, the results of computed jobs are collected by the master computer and inserted into the optimization algorithm for further processing and generating new designs for the next optimization loop. In this example the exchange of input and output data is similar to outer loop optimization (cf. Sect. 4.1.2) by generating and reading files, which are stored on a network drive, accessible to all computers.

For the management of this distributed computing, appropriate software or generated code is required, which must fulfill the following functions:

– Write/modify input-files for the particular type of solver on node computers
– Identify status of nodes in cluster: busy, standby, results available, etc.
– Distribute and launch individual jobs in the cluster
– If necessary, copy and offer additional files required for solving jobs on nodes
– Collect available results and prepare for further processing in optimization algorithm
– Perform error management: error in result, no response of node (time out), handle loss of node in cluster, etc.

It should be clear that, with this organization (distributing jobs, writing/reading of data for the exchange, and waiting time) in addition to the optimization algorithm runtime and the computation of the designs, parallelization includes an extra effort. Compared to the time saved overall, this time cost, the so-called overhead, should be small. It is obvious that parallelization is more efficient as the computation time for one individual design increases. On the other hand, if individual designs can be computed quickly, it is possible that the overhead causes even slower process times when using parallelization than in a common sequential procedure on only one workstation. In Fig. 3.8 we can see the speed-up for problems with different computation times when using parallelization. For each problem (small, medium,

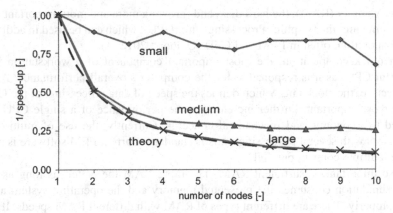

Fig. 3.8 Parallelization speed-up for problems with different computation time

large), we compute 50 variants with a various number of nodes. Furthermore, the theoretical maximum achievable speed-up is indicated, without the time cost for the distribution of jobs.

Efficient computer clusters have all nodes at full capacity and without unnecessary waiting times. For example, if there are 10 parallel tasks to be processed, it would be absurd to use nine nodes in the cluster with eight of them waiting 50 % of the time. With 5 nodes we can reach the same decrease of process time, but using 10 nodes would be ideal. We need to coordinate the number of nodes with the number of jobs we can compute in parallel. Furthermore, the performance of all nodes in the cluster should be comparable to prevent bottlenecks due to diverse computation times of design evaluation.

The relative speed-up S_p of parallel jobs can be calculated by

$$S_p = \frac{T_1}{T_p},$$

where T_1 is the time needed for the sequential process on one computer and T_p is the time when using p computers in a cluster to parallelize jobs.

How many nodes we use in a cluster depends on not only the number of available nodes, but also on the specific optimization task. When using commercial solvers, the number of available software licenses can also limit the number of nodes. Currently, most FEM software providers offer particular license packages for parallel computing.

Benefit of Hardware Raising

Another way to reduce computation time and accelerate the optimization process is the use of well-equipped workstations with appropriate hardware. Here we focus especially on the performance of the Central Processing Unit (CPU), the Random-

Access Memory (RAM), the hard disk, and (growing more and more important and very promising) the Graphics Processing Unit (GPU), which can be used in addition for computation, often in FEM and CFD (Ohlhorst 2012).

For quick computation, the most important component of a workstation is a powerful CPU, as it is responsible for the computer's overall performance. After architecture, the clock rate, which defines the speed of data processing, of the CPU is the most important. Further increases to the performance of a single CPU are limited by economic and mechanical concerns. Currently, the use of multi-core processors with several processing units is standard. Current FEM software is able to use multiple cores in parallel.

Also important, is sufficient RAM. The more RAM the better, as long as the FEM simulation consumes the presented memory and the operating system allocates properly. There are different types of RAM with different RAM speeds. If the system, especially the motherboard, can be upgraded to faster RAM, this will enhance performance, too.

The massive amount of data in an FEM simulation must be handled by large hard drives. To prevent a bottleneck when using high performance CPUs, it is important that the storage component offers sufficient read and write capability. Today Solid-State-Drives (SSD) provide excellent properties for fast processes.

The newest, very efficient strategy for hardware acceleration is the inclusion of high performance GPUs in general-purpose computing. CPUs are developed for universal use and sequential processing. In contrast, the architecture of GPUs is designed for massive parallel processing. In FEM, we can release the equation solving part, which takes about 70 % of the total evaluation time, to the highly efficient GPUs (Güttler 2014). To take advantage of increased process speed by using GPUs, there are special license packages offered by commercial software suppliers.

In general, it is important when buying or upgrading a workstation to verify all components of the system to avoid bottlenecks. All parts must interact well to guarantee a high efficiency and enable fast computation processes.

Conclusions

With the emergence of commercial software for optimization, sensitivity studies, or the evaluation of robustness and reliability, the information of many design variants can be considered and handled. Often computation time limits such detailed investigations, especially when we have complex multi-physics simulations. Currently, parallelization and hardware acceleration with GPUs is a common tool to accelerate such time consuming studies.

But when applying the optimization methods mentioned in this book, the correct strategy, previous experience in optimization, and good preparation are also important for a quick route to the optimal design. Optimization is not simply pressing the start button and waiting for a result. We need to study and understand the problem, find additional information about its parameters and reuse this for the next optimization steps. Well thought-out action is more significant than computing a lot of different variants. Time saving also entails preventing failures or wrong definitions that lead to meaningless results in optimization. However, if an optimization run is canceled, it is beneficial to have access to the results obtained in the partial run to identify and resolve conflicts for further studies.

References

Berke, L., Patnaik, S. N., & Murthy, P. L. N. (1993). Optimum design of aerospace structural components using neural networks. *Computers and Structures, 48*, 1001–1010.

Coelho, L. D. S., & Mariani, V. C. (2006). Particle swarm optimization with Quasi-Newton local search for solving economic dispatch problem. *IEEE International Conference on Systems, Man and Cybernetics, 4*, 3109–3113.

Güttler, H. (2014). *TFLOP performance for ANSYS mechanical 2.0*. Nürnberg: ANSYS Conference 32. CADFEM Users Meeting.

Lagaros, N. D., & Papadrakakis, M. (2004). Learning improvement of neural networks used in structural optimization. *Advances in Engineering Software, 35*, 9–25.

Ohlhorst, F. J. (2012, December). Optimize workstations for faster simulations. In *Desktop Engineering – Technology for Design Engineering* (pp. 68–70). Retrieved June 11, 2015, from http://www.engineerquiz.com/issues/level5_desktopengineering_201212.pdf

Plevris, V., & Papadrakakis, M. (2011). A hybrid particle swarm—gradient algorithm for global structural optimization. *Computer-Aided Civil and Infrastructure Engineering, 26*, 48–68.

Rechenberg, I. (1994). *Evolutionsstrategie '94*. Stuttgart: Frommann-Holzboog.

Steinbuch, R. (2010). Successful application of evolutionary algorithms in engineering design. *Journal of Bionic Engineering, 7*(Suppl), 199–211.

Surjanovic, S., & Bingham, D. (2015). Virtual library of simulation experiments: Test functions and datasets. Retrieved May 12, 2015, from http://www.sfu.ca/~ssurjano

Widmann, Ch. (2012). *Strukturoptimierung mit Neuronalen Netzen*. Master thesis, Reutlingen University.

Chapter 4
Application to CAE Systems

Rolf Steinbuch, Andreas Fasold-Schmid, Simon Gekeler, and Dmitrii Burovikhin

Due to the broad acceptance of CAD-systems based on 3D solids, the geometric data of all common CAE (Computer-Aided Engineering) software, at least in mechanical engineering, are based on these solids. We use solid models, where the space filled by material is defined in a simple and easily useable way. Solid models allow for the development of automated meshers that transform solid volumes into finite elements. Even after some unacceptable initial trials, users are able to generate meshes of non-trivial geometries within minutes to hours, instead of days or weeks. Once meshing had no longer been the cost limiting factor of finite element studies, numerical simulation became a tool for smaller industries as well.

In the early days of automated meshing development, there were discussions over the use of tetragonal (Fig. 4.1) or hexagonal based meshes. But, after a short period of time, it became evident, that there were and will always be many problems using automated meshers to generate hexagonal elements. So today nearly all automated 3D-meshing systems use tetragonal elements.

Another question arose about the adaption of the elements during a convergence study of an analysis to the local geometry. Most systems relied on h-adaptivity, which is based on the refinement of the mesh in regions of high stress concentration (Fig. 4.2). Only a few systems proposed to use p-adaptivity, which increases the degree of the polynomial interpolation of the element edges (Fig. 4.3). Today we observe that most automated meshing and adapting systems in 3D use tetra 10 elements, a parabolic interpolation of the edges and h-adaption, a refinement of the elements.

R. Steinbuch (✉) • A. Fasold-Schmid • S. Gekeler • D. Burovikhin
Hochschule Reutlingen, Reutlingen Research Institute, Alteburgstraße 150, 72762 Reutlingen, Germany
e-mail: Rolf.Steinbuch@Reutlingen-University.DE; Andreas.Fasold-Schmid@Student.
Reutlingen-University.DE; Simon.Gekeler@Reutlingen-University.DE; Dmitrii.
Burovikhin@Reutlingen-University.DE

© Springer-Verlag Berlin Heidelberg 2016
R. Steinbuch, S. Gekeler (eds.), *Bionic Optimization in Structural Design*,
DOI 10.1007/978-3-662-46596-7_4

Fig. 4.1 FE-Mesh of a 3D
geometry using automated
meshing and TETRA10
elements

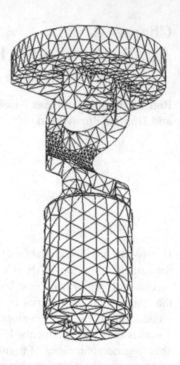

Fig. 4.2 Local h-adaptive
refining of meshes near
critical segments of the part

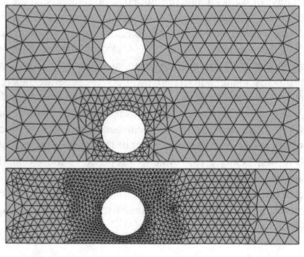

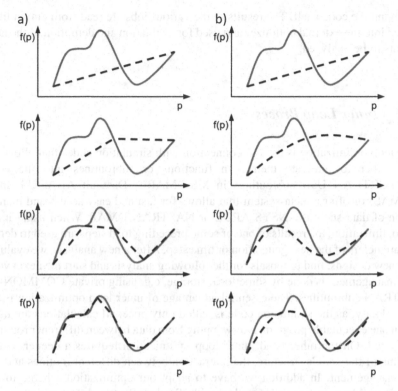

Fig. 4.3 Comparing h and p-adaptive approximation of a curve. (**a**) **h-adaptive**: use more intervals. (**b**) **p-adaptive**: use higher order polynomials

4.1 Inner and Outer Loop Optimization

Andreas Fasold-Schmid, Simon Gekeler, and Rolf Steinbuch

When doing optimizations, the use of commercial simulation software, typically a Finite Element code or a CAE-System with a more or less open simulation subsystem is recommended, as it is not too easy to code a reliable simulation package on one's own.

Optimization then means to start a series of jobs of various different variants, evaluate them, analyze the optimization's step success and define the next set of variants or designs to be studied. This cycle has to be repeated until a predefined stop criterion is met. This stop criterion may be a maximum number of generations, a level of the goal or no further improvement of the goal during some cycles.

To realize such an optimization using a given code, two different approaches are used: inner and outer loop optimization. The first one is integrated into the code as part of its internal structure and data management. All variants, evaluations and definitions of new designs are done within the code. The outer loop optimization works more or less as an editor that modifies the input sent to the code without modifying the code itself. The results of the various jobs are read from output files, entered into the external optimizer and used for evaluation and definition of the next variants to be analyzed.

4.1.1 Inner Loop Process

Inner loop optimization is used in connection with simulation codes that allow the users either to integrate their own functions or subroutines into the code (e.g. FORTRAN User subroutines in MSC-MARC, Dassault ABAQUS, Intes PERMAS) or offer a meta-system that allows for fast and easy access and manipulation of data such as ANSYS APDL or NASTRAN DMAP. When using inner loop optimization, the results of one or some preceding load steps are used to define the parameters of the next generation or time step. After a new analysis, we evaluate these new designs, find proposals for the following analysis and start the next cycle. Data management is done by some local storage, e.g. using private COMMONs in FORTRAN subroutines. The essential advantage of inner loop optimization is the high velocity, as the simulation code is called only once, all calculations are done within one executable program. No swapping from data between different programs is required. On the other hand, inner loop optimization requires a deeper understanding of the code's structure, the interaction between different routines and the data management. In addition we have to adapt our optimization scheme to the specific problem we are going to deal with. So the effort to define an inner loop optimization may be essentially larger than for the outer loop optimization.

When trying to do an inner loop optimization, the users must check whether all data they need for the optimization are accessible. These data should include the values of the goal and the restrictions and the free parameters they want to modify during the course of the optimization. In most cases the goal and the restrictions are relatively easy to find, as they are the results of the jobs or of different load steps, depending on the way the codes handle the repeated analysis of one problem during the optimization process. The input data, which stand for the free parameters, need to be modified to produce the variants of the studies. This should be not too difficult for material or loading data. For the modification of geometric shapes, unfortunately a new problem arises. We want to modify geometric entities, but in the codes often only the nodes that define the elements are available. Therefore we need to relate the shifting of nodes to the desired change in geometry. This may be not too complicated if the simulation system is based on geometric data and does an automated remeshing for each time step. If only the nodes are available, more care has to be taken.

Example 4.1 The idea of local growth in Sect. 2.6.2 to reduce notch stress by shape smoothing is done by an inner loop optimization. It is a sequence of repeated load steps, where each step uses the same material and loading, but a geometry that is adapted to the local stresses and so remove local stress concentrations. We start the job using a given initial geometry (Fig. 4.4a) which corresponds to a given

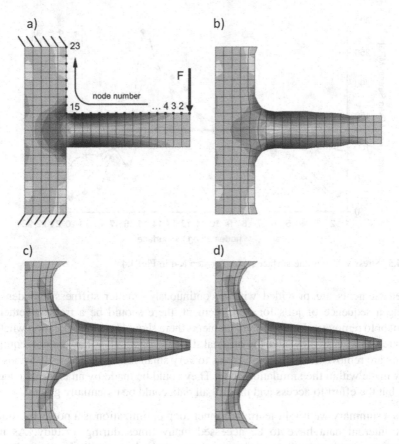

Fig. 4.4 Inner loop optimization: shaping geometry related to a given target stress value. (**a**) Initial geometry. (**b**) Geometry after 9 steps. (**c**) Geometry after 41 steps. (**d**) Geometry after 65 steps

position of the nodes along the surface. Next we read the results of the first load step, e.g. the stresses along the surface we are allowed to modify. For each of these nodes, we decide whether to move to the outside to reduce high stresses or, if we obtain low stresses, we move the nodes inside the geometry. These steps of moving the nodes must not be too large to avoid loss of smoothness. After some steps, a rather nicely shaped contour should be the result of our smoothening process (Fig. 4.4b–d).

The stress distribution at the surface nodes for the steps a-d (Fig. 4.4) is depicted in Fig. 4.5. With proceeding optimization the stresses along the surface converges to the target stress value of 130 MPa.

Example 4.2 The Topological Optimization (Sect. 2.6.1) is mostly handled as an inner loop optimization in commercial codes. We do an initial analysis of the space filled with elements, and decide which one to reduce. In the next steps, these less

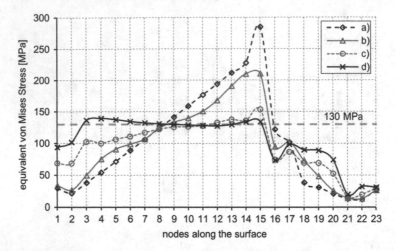

Fig. 4.5 Stress values at the surface nodes for step a–d in Fig. 4.4

loaded elements are provided with a continuously smaller stiffness and density. During a sequence of jobs for each element, there should be a trend whether to keep it or to remove it. Finally for all elements there should be a classification whether they are required, so they have the original stiffness, or whether they are not required, so they are removed, which corresponds to very small stiffness. These decisions are easily made within the simulation code. They could be made by an external code as well, but the effort to access and modify all data could be essentially greater.

As a summary we might learn that inner loop optimization is a powerful tool, if many internal data have to be accessed many times during a study. As it is incorporated within the flow of the simulation code, it generally is relatively fast. But as we have to access very specific data for each problem given, there is some work necessary to establish the procedures. Furthermore some code providers have the tendency to modify their data handling during introduction of new releases. In such cases it may be inevitable to rewrite the whole inner loop process, which might be some undesirable effort.

Table 4.1 in Sect. 4.1.2 offers a comparison of the characteristics of inner and outer loop optimization.

4.1.2 Outer Loop Process

Outer loop optimization is used, if there exists access to the relevant data of a problem by reading output files and editing input files of the models to be analyzed. Just as a human user who tries to find an optimal design and therefore runs series of variants of the initial design, the outer loop optimizers run jobs, vary their parameters and check the performance of the designs. Then any appropriate optimization strategy will be used to find better solutions.

Table 4.1 Comparison of inner and outer loop optimization

Inner loop optimization	Outer loop optimization
Quick data management (intern processed)	Relative slow data management (write/read input-/output files, start CAE-system)
Inflexible: optimization procedure needs to be adapted to CAE-system, costly to do changes	Flexible due to changes in optimization algorithm
Easy data management: internal handling of variables and parameters	Complex data management: external data transfer via input-/output files
Optimization process (if already implemented) available without further preparation	Adaption of generating input files and reading output files for different problems necessary
Optimization algorithm needs to be coded in specific programming language	Optimization algorithm (external optimizer) in optional programming language
Optimization algorithm only for specific CAE-system	Optimization algorithm (external optimizer) available for different CAE-systems
Often the CAE-system offers tools to do optimization postprocessing	Optimization postprocessing needs to be coded in external optimizer

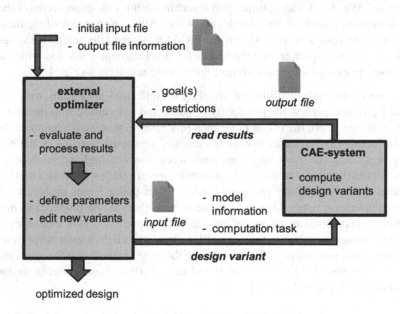

Fig. 4.6 Workflow of outer loop optimization with commercial CAE-system

Many of the commercial CAE-systems with integrated simulation tools offer such access to the problems parameters. Then it is not too difficult to set up a code that does the outer loop optimization. Figure 4.6 shows the program flow of such a code. As preparation, first we need to set up the model we want to optimize in a usual way in the CAE-system. After a computation we obtain the input file and at least one output file. We use the input file, which contains all model information and the defined computation task, as a template for the modification of the relevant parameters to create the design variants. In the output file, we identify the structure of the results and locate data of optimization goals and restrictions. After including this information in the external optimizer, we start the optimization process in which the optimization algorithm defines new design variants as specific input files and calls the CAE-system to compute each job automatically. The quality of each design is evaluated by reading the corresponding output file and, according to this result, new designs will be generated until the optimization algorithm converges or will be stopped.

The coupled structure of an outer loop process makes the optimization procedure itself independent from CAE-systems and thus it is flexible for improvements or quick variations in the optimization strategy. Once the optimization algorithms are established and available in the outer loop framework, using different CAE-systems or different optimization parameters is simply done by adapting the modification of input files and updating the identification of result data in the output files.

Example 4.3 Figure 4.7 represents an excerpt of the input and the output file for one design variant when doing an outer loop optimization of the 10 rods frame problem (cf. Fig. 3.1b). The optimization algorithm defines the cross section values in the input file for each of the 10 rods (Fig. 4.7a). After computation of the design variant with the commercial FEM-solver PERMAS, we read the results for weight, which we want to minimize, and the values for displacement and stress which are not allowed to exceed the limits to meet the defined restrictions (Fig. 4.7b).

Because of the high flexibility of coupling optimization algorithms with commercial CAE-systems and the possibility of quick and easy changes in the optimization procedure itself, the use of outer loop processes in general is recommended. Even including CAD-systems to realize complex geometric changes in the outer loop process may be possible. This framework enables high accessibility on possible optimization parameters. The disadvantage of relative slow data handling compared to the inner loop process vanishes when doing larger problems with longer computation times. Users must take care as there is the potential to make mistakes when defining an automated process for the modification of input files and reading of parameter values in the output files. Problems might appear when, due to the computation task, output files will change their structure during the generation of design variants. A comparison of the characteristics of inner and outer loop optimization is listed in Table 4.1.

4.2 Implementation in CAE-Systems

Dmitrii Burovikhin

Shape optimization is an important part of today's engineering process, enabling us to design mechanical parts and structures with minimal costs and effort. Combined with commercially available CAD software, it gives us a powerful tool for creating efficient CAD models that can be used in production to minimize possible waste, energy consumption and costs.

a)
```
      :
    $STRUCTURE
  !
    $COOR
          1    7.200000E+02     3.600000E+02     .000000E+00
          2    7.200000E+02      .000000E+00     .000000E+00
          3    3.600000E+02     3.600000E+02     .000000E+00
          4    3.600000E+02      .000000E+00     .000000E+00
          5     .000000E+00     3.600000E+02     .000000E+00
          6     .000000E+00      .000000E+00     .000000E+00
    $END STRUCTURE
  !
    $SYSTEM
     $ELPROP
      ALLBARS            MATERIAL = MAT1
      1                  GEODAT = P1
      2                  GEODAT = P2
      3                  GEODAT = P3
      :                      :
      :                      :
  !
     $GEODAT    FLANGE    CONT = SECTION   NODES = ALL
      P1         3.098100E+01
      P2         1.000000E-01
      P3         2.317140E+01
      :              :
```

b)
```
  :
> Element set ELEMDISP with 10 elements
 +--------------------------+--------------------------+
 |        Type              |          Mass            |
 |--------------------------+--------------------------|
 |     Consistent           |       5.062299E+03       |
 +--------------------------+--------------------------+
> Nodal point displacements in component system (Book USR.DISP)
 ----------------------------------------------------------------
 Load-Pattern        1              1              1
    Nodes            u              v              w
 ----------------------------------------------------------------
          1   1.87613E-01  -2.00000E+00   0.00000E+00
          2  -5.36340E-01  -1.99182E+00   0.00000E+00
          3   2.35457E-01  -7.41984E-01   0.00000E+00
          4  -3.06641E-01  -1.64107E+00   0.00000E+00
 ----------------------------------------------------------------

  Elemental Stresses          Load Pattern No.        1


     Node    Uniaxial Stress


          Element No.          1      Type   FLA2
      5    6.5405E+00
      3    6.5405E+00
      :       :                         :
```

Fig. 4.7 Excerpt of input and output file of the 10 rods frame problem in PERMAS. (**a**) Input file defines cross sections of the rods. (**b**) Output file contains resulting weight, node displacements and stress of each rod

The various mentioned optimization algorithms, such as Particle Swarm Optimization and Evolutionary Optimization, or any other strategy, might be integrated in CAD systems such as PTC Creo Parametric, Siemens NX or other CAE-Systems that allow access to parametric models and running simulations with the modified models (Burovikhin 2015). The optimization algorithms used in this chapter have been written as code in C++, and linked to the corresponding sections of the original code or used as overlay (cf. Sect. 4.1). Such integration allows us to optimize single CAD models, as well as entire assemblies to minimize or maximize possible objectives such as the mass (volume or weight) of a mechanical structure, maximum stress and maximum displacement or fit Eigen frequencies to restricted ranges.

The type of optimization we present in this chapter is a mono-objective parametric shape optimization. The term "mono-objective" means that we have only one goal function to be optimized, as opposed to "multi-objective" optimization. Expansion to multi-objective optimizations (cf. Sect. 6.2) may be included without many problems.

4.2.1 Mono-objective Parametric Shape Optimization

The basic terms of optimization have been introduced throughout the other chapters, so we mention here only the specific terms we need for integration into commercial CAE systems. In an unconstrained optimization problem, the area of the search space is defined by the objective function and the set of parameter ranges. However, here we deal with constrained optimization problems, where a defined set of constraints restricts the area of the search space and delimits a restricted subspace to be searched for an optimum solution.

We then search for an optimum by iteratively modifying the dimensions or any other input of a CAD-model of a component or an assembly—the free parameters—thus changing the shape or performance of the part. This is where the term "shape optimization" comes from. As both PTC Creo and NX enable us to create parametric models where every dimension can be a parameter in the optimization process, we easily have access to these parameters.

When it comes to mono-objective optimization, we have only one objective function to optimize. By modifying the free parameters of a part according to a certain optimization algorithm (PSO, EVO or others), we receive a set of so-called responses in return. A response is a property of a part that is specific to a certain combination of the design variables. The list of responses may include the value of mass, volume, weight, the maximum stress, maximum displacement, etc. Out of these responses we can choose our objective function and the constraints. *A priori* an optimization algorithm does not distinguish between the responses. This means that any response can be chosen either as the objective function or as a constraint. Of course, the user should not choose implausible combinations of the responses that have no application in real life.

4.2.2   *Formulation of Structural Optimization Problem*

A general continuous structural optimization problem can be stated as follows:

$$
\begin{aligned}
&\min z(\mathbf{p}), \quad \mathbf{p} = [\mathbf{p}_1, \dots, \mathbf{p}_n]^T, \quad \mathbf{p} \in \mathbb{R}^n \\
&\text{subject to } g_k(\mathbf{p}) \leq 0, \quad k = 1, \dots, m \\
&\qquad\qquad \mathbf{g} = [g_1, \dots, g_m]^T, \\
&\qquad \mathbf{p}_i^L \leq \mathbf{p}_i \leq \mathbf{p}_i^U, \quad i = 1 \dots n
\end{aligned}
\tag{4.1}
$$

where $\mathbf{p}$ is a vector of length n containing the design variables, $z(\mathbf{p}) : \mathbb{R}^n \rightarrow \mathbb{R}$ is the objective function, which returns a scalar value to be minimized (for example, the weight of the structure), the vector function $\mathbf{g}(\mathbf{p}) : \mathbb{R}^n \rightarrow \mathbb{R}^m$ returns a vector of length m containing the values of the inequality constraints evaluated at $\mathbf{p}$, and $\mathbf{p}^L$, $\mathbf{p}^U$ are two vectors of length n containing the lower and upper bounds of the design variables, respectively. The above formulation contains only inequality constraints. Equality constraints may be used in structural optimization as well. Generally they tend to reduce the size of the parameter space.

A typical constraint k in structural optimization has the form

$$
g_k(\mathbf{p}) = q_k(\mathbf{p}) - q_{allow,k},
\tag{4.2}
$$

where $q_k(\mathbf{p})$ is a response measure (typically stress or displacement) for design $\mathbf{p}$ and $q_{allow,k}$ is its maximum allowable absolute value. It should be noted that $q_k(\mathbf{p})$ is often taken as the maximum (worst) value of the corresponding response measure among all nodes or elements of the model (Plevris and Papadrakakis 2011).

For example, if the k-th constraint is a stress constraint of the type $|\sigma| \leq \sigma_{allow}$ that applies for all N_e model elements, then for this constraint a single response measure is calculated as

$$
q_k(\mathbf{p}) = \max_{i=1}^{N_e} \{|\sigma_i|\}.
\tag{4.3}
$$

To explain the implementation of the optimization into commercial codes, we give examples of two commercial CAE-systems which are widely used in mechanical engineering. This does not imply any preference for this CAE-Systems or rejection of others, where similar ideas might be realized as well. We just want to indicate that it is possible to do it, and to motivate other users to check, how to implement comparable optimizers in their CAE-systems.

4.2.3   Bionic Parametric Shape Optimization with PTC

PTC Creo is a parametric, integrated 3D CAD/CAM/CAE solution created by Parametric Technology Corporation (PTC). It was the first to market with parametric, feature-based, associative solid modeling software.

Trail Files

When it comes to automating PTC Creo jobs, a trail file is an integrated part of the process. The trail file is a record of all menu choices, dialog-box choices, selections, and keyboard entries for a particular interactive PTC Creo session. With trail files, we can view the record of activity so that we can reconstruct a previous working session or can recover from an abruptly terminated session. Trail files are editable text files.

When we run a trail file, all the selections are replayed in the exact original order in which they were made. Before running a trail file, we should rename it to avoid confusion in the data sets.

If PTC Creo Parametric crashes, we have to edit the trail file up to the line where the crash occurs or we will simply repeat all the commands leading up to the crash and the crash itself. We can use the trail file feature to automate a series of actions and create a routine that can be repeated over and over again every time we run the trail file. In the framework of this chapter, we need to create a routine that updates our part every time the free parameters are modified and then computes the value of the objective function and the values of the constraints for the current particle (PSO) or individual (EVO).

Running PTC Creo Parametric in a Batch Mode

If we want our application to perform operations on PTC Creo Parametric objects without any user interaction or without Graphical User Interface (GUI), we can run PTC Creo Parametric in batch mode. A useful technique when designing a batch-mode application is to use command-line arguments to PTC Creo Parametric as a way of signaling the batch mode and passing in the name of a batch control file. A batch-mode operation should run without displaying any graphics. We can use trail files to script the batch processing as it is shown above.

Example 4.4 Let's consider a simple example: we have an I-beam made of steel, with a fixed length $l = 400$ mm. The beam is fixed at one end, and there is a distributed force $F = 500$ kN applied to the upper flange (Fig. 4.8). The I-beam has six parameters (the dimensions of the cross-section) which can be modified within certain ranges:

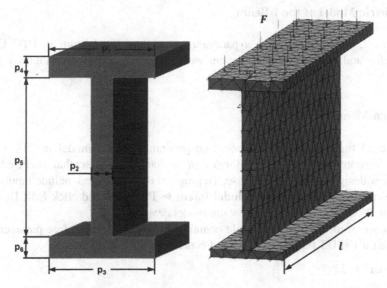

Fig. 4.8 Dimensions of the I-beam used as free parameters for optimization

- $p_1 \in [90, 150]$
- $p_2 \in [20, 60]$
- $p_3 \in [90, 150]$
- $p_4 \in [10, 30]$
- $p_5 \in [150, 200]$
- $p_6 \in [10, 30]$

We also have restrictions for the maximum acceptable values of stress and displacement

$$|u| < u_{max}, |\sigma| < \sigma_{max}.$$

Our goal is to find minimum mass of the I-beam (the objective function), keeping maximum stress below 360 MPa and maximum displacement smaller than 1 mm. We create a procedure within the optimization code that returns the values of the mass, stress, and displacement. All three values depend on the set of the free parameters. These parameters are generated during the simulation run and then passed to our procedure by the optimization algorithm. Then the procedure exports these parameters into an external text file which is used by PTC Creo Parametric to regenerate the model. When new values of the free parameters are exported into the text file, the procedure opens a new session of PTC Creo Parametric where the model is regenerated according to the values of the free parameters, and all the necessary calculations are carried out. After it is done, PTC Creo Parametric creates a folder containing the simulation results. The procedure extracts the values of maximum stress and maximum displacement from one of the files in the folder and passes it back to optimization algorithm.

Parametric Model of the I-Beam

First of all, we need to build a parametric model of the I-beam in PTC Creo Parametric and name all the dimensions we want to use as design variables.

Program Menu

To proceed further, the model should be programed. Each model in PTC Creo Parametric contains a listing of major design steps and parameters that can be edited when new design specifications arise. To program the model and include inputs for the dimensions go to **Model ▶ Model Intent ▶ Program** and click **Edit Design** from the PROGRAM menu to view the model design.

Now we need to create a text file containing the values of the free parameters. Let's call it INPUT.txt. This file should contain the following data:

Parameter1 $= 20$
Parameter2 $= 53$
. . .
ParameterN $= 60$

This file will be modified by the optimization code and then used to update the model. To create a trail file for this particular problem, open PTC Creo Parametric (if we have it open already, exit and launch it again), reopen the model and regenerate it. PTC Creo will ask us if we want to read a file. Select the INPUT.txt file we've created and the model will update. Perform a simulation study and exit PTC Creo.

Copy the trail file that has been created during this session (we should be able to find it in our working directory) under a new name with the extension *.txt. Let's take a look at the whole process one more time. We have a code for a specific optimization algorithm (PSO or EVO). The code runs PTC Creo in a batch mode to obtain the values of the mass, maximum stress and maximum displacement that correspond to the current values of the free parameters. The values of the free parameters are generated by the optimization algorithm and exported into a text file (1) (cf. Fig. 4.9). Then a new session of PTC Creo Parametric is launched using a trail file as an argument (2) meaning that all the commands in the trail file will be carried out, such as opening of the model (3), regeneration of the model according to the text file (4), simulation study (5) and creation of the results folder (6). After that, the values of the maximum stress and displacement are extracted from one of the result files (7).

The optimization results are shown in Fig. 4.10. We observe here that, with the same number of objective function evaluations, EVO and PSO show similar performance in the convergence of the best mass value. The accuracy of the final solution also depends on the initial settings for the algorithms. The choice of the settings depends on the problem at hand (cf. Sect. 3.1).

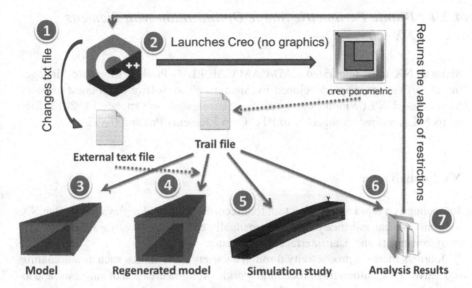

Fig. 4.9 General concept of optimization using PTC Creo

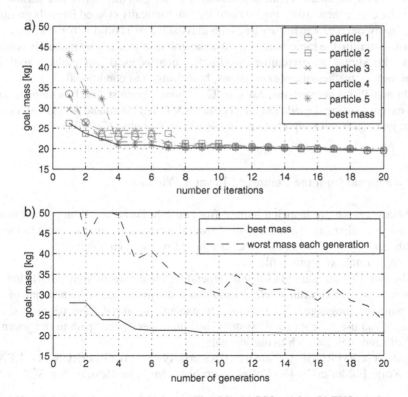

Fig. 4.10 Minimizing the mass of the beam (Fig. 4.8). (**a**) PSO results. (**b**) EVO results

4.2.4   Bionic Parametric Shape Optimization with Siemens NX 9.0

Siemens NX is an integrated CAD/CAM/CAE/PLM (Product Lifecycle Management) software package developed by Siemens PLM Software. Its latest version, used here—NX9 (Windows x64 only)—was released on October 14, 2013. Siemens NX is a direct competitor to PTC Creo Elements/Pro and CATIA.

NX Journaling

Journaling is a rapid automation tool that records, edits, and replays interactive NX sessions. We can enhance journals by manually editing them with simple programming constructs and user interface components.

Journals increase productivity through a variety of scenarios such as automating repetitive tasks, automating procedural workflow, rapid creation of autotests, and as an aid for creating more advanced automation programs.

We can automatically create journals by recording an interactive NX session, or manually create them using any text editor. Automatically created journals produce the same model data and history tree with playback as was originally created during record, assuming the same start state. We can enhance journals by manually editing them with simple programming constructs, such as local and global variables, loops, arrays, mathematical expressions, branching, and conditionals.

On non-Windows systems, Java, or C++ journals must first be compiled and then executed using the **File** tab >**Execute**>**NX Open** command. Replay of Visual Basic journals is not supported on non-Windows systems (NX 9.0).

Run a Journal from the Command Prompt Window

On Windows, we can launch a journal from an NX Command Prompt shell using the run_journal utility. To run a journal from the command line, we need to locate the file called "run_journal.exe" and use the following command:

"run_journal.exe journal_file.vb"

The path to "run_journal.exe" will depend on the details of your installation. In the same way, we can run an NX journal from within a code written in one of the programming languages. For example, to execute a journal from within a C++ code, we can use the command "system" and then specify the "path to run_journal. exe" file and then the path to our journal:

system("c:\crPROGRA~1\crSIEMENS\crNX9~1.0\crUGII\crRUN_JO~1.EXE e:\crWork_Folder\crC++\crCubesAssembly\crModel\crResults_for_SOL_106. vb")

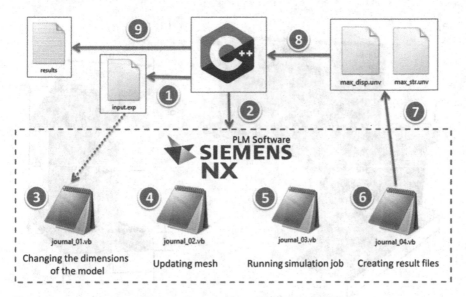

Fig. 4.11   General concept of Bionic Optimization with Siemens NX9.0

General Concept of the Outer Loop Optimization with Siemens NX 9.0

As the term "outer loop optimization" (compare Sect. 4.1.2) implies, our optimization algorithm has two loops—the inner loop and the outer loop. The outer loop is the optimization code itself where the inner loop involves all the operations performed by using the modelling update and the simulation part of the external CAD software. As the process progresses, it launches interactive NX sessions to find the values of the goal function and the constraints for the current particle (PSO) or individual (EVO) by running simulation analysis in a batch mode (Fig. 4.11).

The values of the free parameters are generated by the optimization algorithm and exported into a text file (1) (cf. Fig. 4.11). Then a new session of NX is launched using journals (2). All the commands in the journals will be carried out, such as changing the dimensions of the model (3), updating the mesh (4), simulation study (5) and creation of the result files containing values of stress and displacement (6). After that, the values of the maximum stress and displacement are extracted from the result files (7) and passed back to the optimization code (8). At the end of the optimization process, the final optimization results are exported into the main result file (9).

Example 4.5   To understand how it all works, let us consider the example visualized in Fig. 4.12. Let us assume that we have two I-beams, made of steel, in contact with each other, and we want to minimize the mass of the structure (the objective function) by modifying some of the dimensions (free parameters). Both I-beams, each with a length of 400 mm, are fixed at one end, and there is a distributed force

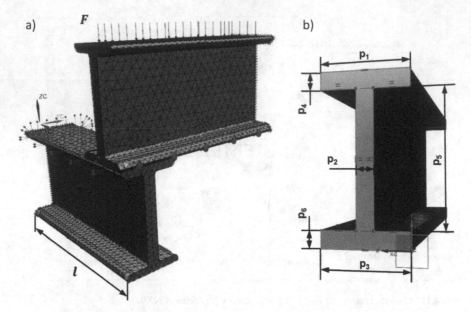

Fig. 4.12 Two beams in contact as an optimization problem. (**a**) Assembly of two beams in contact. (**b**) Initial dimensions of the beams

$F = 100$ kN applied to the upper one. We assume that the maximum stress should not exceed 300 MPa and that the maximum displacement should not exceed 3 mm (optimization constraints).

In Fig. 4.12 we see the dimensions we use as optimization parameters for both I-beams. We modify them within the ranges:

- $p_1, p_7 \in [90, 150]$;
- $p_2, p_8 \in [20, 60]$;
- $p_3, p_9 \in [90, 150]$;
- $p_4, p_{10} \in [10, 30]$;
- $p_5, p_{11} \in [150, 200]$;
- $p_6, p_{12} \in [10, 30]$.

Creating a Parametric Model of the I-Beam

First of all, we need to create a parametric model of the I-beam. It includes creating a set of the free parameters that are associated with the dimensions of the part we want to modify. We also need to assign a material to the part here.

Exporting the Value of Mass of a Part or an Assembly

To be able to obtain the value of mass for every particle (PSO) or individual (EVO), we need to include it in the list of free parameters. First, open the part (or the assembly file). Go to **Analysis** > **Measure Bodies** and select the model (or the entire assembly). Click OK. Go to **Tools** > **Expressions** in the drop-down menu under **Listed Expressions** select **Measurements**. We will see the list of parameters associated with the measurements we've just created. We can rename these parameters.

Note that if our part or assembly has been modified and we want to update our measurements, go to **Part Navigator** and in **Model History** folder we will find **Body Measurement (#)**. Right-click on it, and choose **Edit Parameters**. Then, in the popup window, click OK. Our measurements in **Expressions** dialog box should be up to date. Now we can export these parameters into an external file. Go to **Tools** > **Expressions** and click **Export Expressions to File**. The exported file is generated by the system and ends with a *.exp file extension. If we need to examine or edit the file, we can refer to the format rules shown below (NX 9.0).

Creating a FEM Model of the I-Beam

Go to **File** > **Advanced Simulation**. In the **Simulation Navigator**, right-click on the part and choose **New FEM**... Save the FEM file. Now we can create a mesh.

Creating a SIM Model of the I-Beam

The next step is the SIM file. In the **Simulation Navigator**, right-click on the FEM-model and choose **New Simulation**... Save the SIM file. Now we can apply loads and constraints to the part and run a simulation. We will need four journals to automate the job in NX. To execute the first journal, we need our *.exp file that we have created earlier. Let's name this file "input.exp". The purpose of the first journal is to update the part according to the data in the expression file, compute the mass of the part and export this value into another expression file—let's call it "mass.exp". The following sequence of commands is executed to record the first journal (NX 9.0):

1. Start recording
2. Open the part.
3. Go to **Tools** > **Expressions** and click **Import Expressions from File**. Specify the path to the expression file "input.exp".
4. Go to **Part Navigator**, in the **Model History** folder right-click on **Body Measurement** and in the popup menu select **Edit Parameters**. Click OK in the popup window.

5. Go to **Tools** > **Expressions** and click **Export Expressions to File**. Specify the path to the expression file "mass.exp".
6. Save the changes and close the part.
7. Stop recording.

Since the geometry of the part has been modified by the first journal, we have to update the mesh. This is where we need the second journal. The following sequence of commands is executed to record the second journal:

8. Start recording
9. Open the FEM file.
10. Go to **Simulation Navigator**, right-click on the part file and select **Load**.
11. Go to **Simulation Navigator**, right-click on the FEM file and, in the popup menu, select **Update**.
12. Save the changes and close the FEM file.
13. Stop recording.

The third journal opens the simulation file and runs the simulation job. The following sequence of commands is executed to record the third journal:

14. Start recording
15. Open the SIM file.
16. Click **Solve**.
17. Close the SIM file
18. Stop recording.

The fourth journal creates the result files from which we can extract the values of stress and displacement—our constraints. The following sequence of commands is executed to record the fourth journal:

19. Start recording
20. Open the SIM file
21. In **Simulation Navigator**, go to the **Results** folder of our current solution. Right-click on the **Imported Results** folder and select **Import Results**. Specify the path to the binary result file with the extension .op2 that was previously created during simulation analysis.
22. Go to the **Results** tab on the main toolbar and select the command called **Envelope**. Select the type of results we want to output and create two result files—one for the values of stress and another for the values of displacement.
23. Close the SIM file.
24. Stop recording.

Now we can just execute these four journals one by one every time we need to obtain the value of the objective function and the values of the constraints for a particle (PSO) or an individual (EVO). Figure 4.13 shows optimization results for PSO and EVO.

As we can see, again PSO and EVO perform similar. Both optimization methods show a decreasing mass value until the maximum number of iterations or

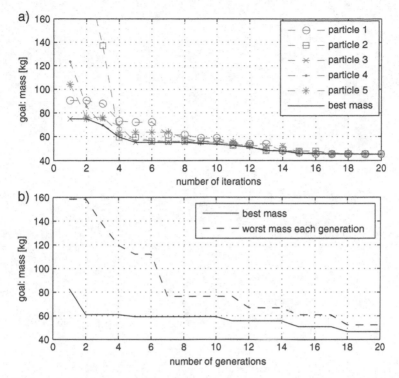

Fig. 4.13 Optimization results of the two beam-contact problem using EVO and PSO. (**a**) PSO results. (**b**) EVO results

generations is reached. Continuing the process or switching to gradient methods may provide further improvement. As we mentioned in Sect. 3.3.1, the convergence rate and the time of execution for both algorithms can be improved by tuning the initial settings.

References

Burovikhin, D. (2015). *Multibody parametric shape optimization: Integration of CAD systems into optimization algorithms*. Master thesis, Reutlingen University.

NX 9.0. *NX Documentation (Help)*. NX 9.0. Retrieved September 12, 2015, from http://www.plm.automation.siemens.com

Plevris, V., & Papadrakakis, M. (2011). A hybrid particle swarm—gradient algorithm for global structural optimization. *Computer-Aided Civil and Infrastructure Engineering, 26*, 48–68.

Chapter 5
Application of Bionic Optimization

Rolf Steinbuch, Iryna Kmitina, Tatiana Popova, Simon Gekeler,
Oskar Glück, and Ashish Srivastava

To illustrate the power and the pitfalls of Bionic Optimization, we will show some examples spanning classes of applications and employing various strategies. These applications cover a broad range of engineering tasks. Nevertheless, there is no guarantee that our experiences and our examples will be sufficient to deal with all questions and issues in a comprehensive way. As general rule it might be stated, that for each class of problems, novices should begin with a learning phase. So, in this introductory phase, we use simple and quick examples, e.g., using small FE-models, linear load cases, short time intervals and simple material models. Here beginners within the Bionic Optimization community can learn which parameter combinations to use. In Sect. 3.3 we discuss strategies for optimization study acceleration. Making use of these parameters as starting points is one way to set the specific ranges, e.g., number of parents and kids, crossing, mutation radii and, numbers of generations. On the other hand, these trial runs will doubtless indicate that Bionic Optimization needs large numbers of individual designs, and considerable time and computing power. We recommend investing enough time preparing each task in order to avoid the frustration should large jobs fail after long calculation times.

As a general rule it must be stated, that if time is limited and a critical element, then no automated optimization should be done. All optimization, deterministic or random based, takes time. This time cannot be reduced, even if under the pressure of unrealistic deadlines. If your management thinks they know better, we recommend suggesting that they develop the optimal design themselves.

R. Steinbuch (✉) • I. Kmitina • T. Popova • S. Gekeler • O. Glück • A. Srivastava
Hochschule Reutlingen, Reutlingen Research Institute, Alteburgstraße 150, 72762 Reutlingen, Germany
e-mail: Rolf.Steinbuch@Reutlingen-University.DE; Iryna.Kmitina@Reutlingen-University. DE; Tatiana.Popova@Reutlingen-University.DE; Simon.Gekeler@Reutlingen-University. DE; Oskar.Glueck@Student.Reutlingen-University.DE; Ashish.Srivastava@Reutlingen-University.DE

© Springer-Verlag Berlin Heidelberg 2016   101
R. Steinbuch, S. Gekeler (eds.), *Bionic Optimization in Structural Design*,
DOI 10.1007/978-3-662-46596-7_5

5.1   Earthquake Stability and Tuned Mass Dampers

Rolf Steinbuch

Among the disasters people are exposed to, earthquakes are among the most damaging. Large numbers of fatalities and significant economic losses are reported in many cases. In consequence, engineers have tried in modern times to design buildings that can withstand the dynamic forces or at least do not incur extensive damage. The local physical impact of a seismic event may be modeled as a series of horizontal and vertical shock waves that excite the base of a building. The building's dynamic response may cause severe damage or even the total collapse of the structure. The dynamic response of tall structures may also be increased by the large deformability of the building. To prevent such destruction, various approaches are available, and these may be classified into three strategies: increase the static strength, isolate from the excited ground, or compensate by elastically coupled masses spaced along the buildings' height.

Tuned Mass Dampers (TMD) are one of several methods used to reduce the impact of an earthquake on buildings. They do so by dampening the impact with sets of masses, springs and dampers. Such a damping system with many degrees of freedom may have many local maxima, thus, the optimization of a set of TMD's could be a non-trivial task. Bionic strategies like Evolutionary Strategy (EVO) or Particle Swarm Optimization (PSO) may help to cover larger regions of the parameter space and increase the probability of finding good values, perhaps even the best ones. More details about this application may be found in Steinbuch (2011).

5.1.1   Earthquake and Design for Earthquake Loading

As the mass of the surrounding and shaking ground is essentially larger than the building's mass, the excitation may be considered as displacement controlled. The Eigen frequencies, the Eigen forms and the amplitudes at a given excitation of a dynamic system may be changed by adding new masses and springs. Figure 5.1a presents acceleration records of the Kobe-Earthquake (Berkeley 2011). The dominant frequencies are in the region of 1 Hz and last for seconds. Accelerations, measured in different events, range to values $a = g = 10$ m/s^2 or more (Fig. 5.1b).

As there are no specific types of earthquakes exclusively related to specific places on Earth, earthquake protection in highly active seismic zones has to withstand all typical types of excitation in magnitude and duration that can be expected in the specific zone. We decompose the impact in displacement terms of the ground u_g into its k frequency components (Fig. 5.1b).

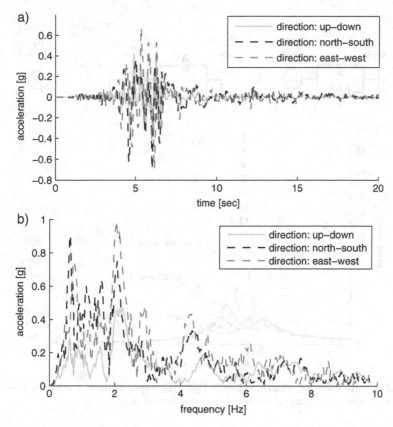

Fig. 5.1 Earthquake: acceleration measurement and spectrum of Kobe earthquake 1995. (**a**) Measured accelerations. (**b**) Spectrum of accelerations

$$u_g = \sum_k u_{gk} e^{ik\omega t} \qquad (5.1)$$

Then we check the impact of each frequency on the dynamic system. As one result, we have the response and sum up the elastic energy of the volume of the structure.

5.1.2   Brief Introduction to Tuned Mass Dampers

Tuned Mass Dampers (TMD), absorbers and compensators are terms used interchangeably for a class of systems that reduce the dynamic impact on structures. To understand the premise of TMD, we start with a single mass oscillator. It is defined by the mass m_1, the stiffness k_1 and the damping c_1 (Fig. 5.2a). Stiffness and damping are attached to some base which may be fixed in time ($u_g = 0$) or has

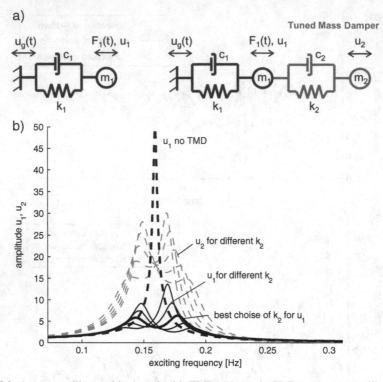

Fig. 5.2 1-mass oscillator without and with TMD. (**a**) Apply TMD to 1-mass oscillator. (**b**) Dynamic response when varying stiffness k_2 to get minimum of u_1

defined displacements ($u_g = u_g(t)$) in time. The system is excited by either the ground motion $u_g(t)$ or a force $F_1(t)$ acting on the mass m_1. The resulting displacement history $u_1(t)$ may be found by integrating the ODE

$$m_1 \ddot{u}_1 + c_1 \dot{u}_1 + k_1 u_1 = F_1(t). \tag{5.2}$$

The well-known solution for the amplitude vs. frequency ω is given by

$$|u_0(\omega)| = \left| \frac{F_1/m_1}{\omega_0^2 - \omega^2 + 2i\gamma\omega} \right|. \tag{5.3}$$

To avoid the large amplitudes near the Eigen frequency, we introduce a TMD (Fig. 5.2a).

$$\begin{pmatrix} m_1 & 0 \\ 0 & m_2 \end{pmatrix}\begin{pmatrix} \ddot{u}_1 \\ \ddot{u}_2 \end{pmatrix} + \begin{pmatrix} c_1 + c_2 & -c_2 \\ -c_2 & c_2 \end{pmatrix}\begin{pmatrix} \dot{u}_1 \\ \dot{u}_2 \end{pmatrix} + \begin{pmatrix} k_1 + k_2 & -k_2 \\ -k_2 & k_2 \end{pmatrix}\begin{pmatrix} u_1 \\ u_2 \end{pmatrix} = \begin{pmatrix} F_1 \\ 0 \end{pmatrix}$$

$$(5.4)$$

Equation (5.4) may be used to derive some properties of TMD. The influence of a TMD on the dynamic response of the initial mass may be learned from Fig. 5.2b. The amplitudes "u_1 with TMD" are significantly smaller at the original Eigen frequency, but they have two maxima of a significant fraction of the initial maximum amplitude. The mass of the TMD may be large as about 5–10 % of the mass m_1.

Multi-mass TMD Systems

For the sake of simplicity, we are restricting ourselves to one-dimensional chains of masses, stiffness and TMD excitation at one end of the chain as indicated in Fig. 5.3.

The corresponding ODE-system of the chain without TMD is given by

$$\mathbf{M}\ddot{u} + \mathbf{C}\dot{u} + \mathbf{K}u = \mathbf{F}(t) \tag{5.5}$$

where $\mathbf{M}$ is a diagonal or consistent mass matrix, $\mathbf{C}$ represents an appropriate damping and

$$\mathbf{K} = \begin{pmatrix} k_{12} + k_g & -k_{12} & 0 & \cdots & 0 \\ -k_{12} & k_{12} + k_{23} & -k_{23} & \cdots & 0 \\ 0 & -k_{23} & k_{23} + k_{34} & \cdots & \vdots \\ \vdots & \vdots & \vdots & \ddots & -k_{n-1,n} \\ 0 & 0 & \cdots & -k_{n-1,n} & k_{n-1,n} \end{pmatrix} \tag{5.6}$$

stands for the chain's stiffness matrix including k_g, the stiffness connecting m_1 to the ground. Just as in the previous problem, we are primarily interested in the harmonic response. We take the elastic energy stored in the system as an indication of the potential damage.

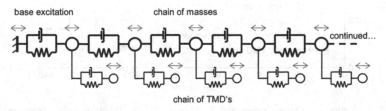

Fig. 5.3 Multimass oscillators

To reduce the dynamic response, we may add some TMDs (cf. Fig. 5.3, where each mass point has its own TMD). Again we use the summed elastic energy stored in the system as an indication of the internally acting destructive load on the main structure.

The optimization problem to minimize the energy in the chain is now multidimensional. We use bionic approaches (cf. Chap. 2) to find efficient designs for the TMDs' mass and stiffness. After some searching of the space of possible values for the m_i, c_i and k_i, we find that the total energy stored in the system may be reduced to about 55 % as indicated in Fig. 5.4 for a 16-mass main system with 16 TMD. The plot shows the fitness function, the elastic energy of the three best parents and the worst parent vs. the generation of the optimization process. We used an Evolutionary Strategy (cf. Sect. 2.1) with 20 parents, 60 kids and a large mutation radius for 100 generations. Figure 5.4 also identifies an additional optimization. Having found an energy level that is close to the minimum of the search using an Evolutionary Strategy after 100 generations, we want to minimize the summed mass of the TMDs. Figure 5.4 indicates that there might be some designs with the same compensation efficiency but essentially smaller masses. The optimization strategy is able to find TMD systems which are nearly as efficient as the best found so far but have a significantly smaller mass—about 20 % of the mass found previously. Now we are able to propose efficient and relatively lightweight designs for TMDs of the 16 masses problem. It is obvious that this approach may be transferred to other dynamic models with an arbitrary number of masses. The strategy used is one of various methods of a Multi-Objective Optimization

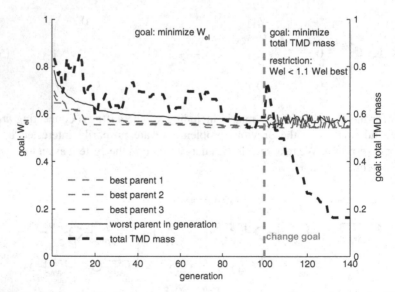

Fig. 5.4 Reduction of elastic energy and TMD-mass for the 16-mass main system using an Evolutionary Strategy

(MOO) in which we seek to optimize energy impact and mass. Section 6.2 discusses more about such MOO.

5.1.3 A Simplified Approach to Study TMD in High Buildings

To come up with quick and realistic proposals of the dimensioning of TMD in high buildings under earthquake impact, we developed a simple beam element representing a segment of a high and slender building including the TMD. Each element is used to model one or more floors of the edifice under consideration. The TMD are an integral part of these elements. The masses, dampers and stiffness associated with these segments are the TMDs' free variables in the optimization process. Every node has 8 degrees of freedom (dof), 3 dof for the nodes' translations, 3 dof for the nodes' rotations and the 2 dof for the in plane displacements of the TMD (Fig. 5.5a). We may use these beam elements to assemble a building by assigning each beam typical stiffness and mass data for the corresponding floors (Fig. 5.5b).

The mass, stiffness and damping of the compensators are used as free parameters for an optimization using this reduced model. The elastic energy stored in the model could be the objective to be minimized, a proposal of the corresponding TMD the result.

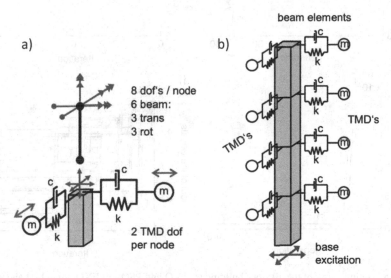

Fig. 5.5 Simple beam-element represents a segment of a building including a TMD. (a) New beam element. (b) The building represented by beam elements

Parameters of the Optimization Process

For the following example, we use a vertical solid steel body with a cross section of $5 \times 6 \text{ m}^2$ and a height of 100 m. We divide this column into 10 floors. The base excitation is given by the Kobe earthquake north-south (Fig. 5.1, Berkeley 2011). Placing three compensators at the floors number 5, 7 and 9 showed very promising responses.

Having decided about the number and position of compensators, we define a range of the dimensions of the compensators' masses, and stiffness. Here, once more, contradictory effects have to be taken into account. Ranges too small prevent the optimization search from finding interesting regions; ranges too large present a danger that the process converges towards infeasible solutions, as, for example the total weight of the compensators exceeds the building's mass.

To compare the efficiency of the strategies, we run the optimization using EVO and PSO, both with 20 particles or kids in each step/generation (Fig. 5.6a, b).

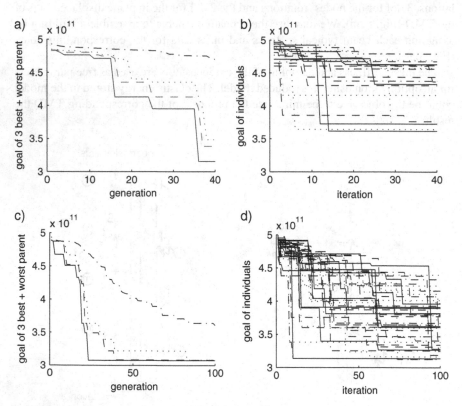

Fig. 5.6 Optimization of the 10 story building by EVO and PSO. (**a**) EVO: three best and worst parents data show slow progress. (**b**) PSO: best individuals seem to stick at local maximum. (**c**) EVO: increasing the number of generations yields better results. (**d**) PSO: no more sticking, better results found, close to the ones of EVO

The tendencies of EVO shown in Fig. 5.6a for the three best and the worst parents indicate that the optimization is far from being finished after the 40 generations. As long as there is a significant span between the parents, there is a potential of further improvement. The PSO (Fig. 5.6b) shows fast improvement in the first steps. Then no further essential decrease of the objective is observed. It seems that the best individual found a local maximum, and that, while some of the other particles tend to join the best particle, we do not find any better solutions.

A new analysis of the same problem, now using 40 parents or particles and 100 generations/steps produces the results shown in Fig. 5.6c, d. The optima found are better than the ones of the 40 generation runs, but here once more PSO is less efficient than EVO. The results ought to motivate one to be concerned about the quality of the results. Without a skeptical review, Bionic Optimization has the potential to lead to weak or untrustworthy statements.

After the preliminary studies shown in Fig. 5.6, we would now decide to use EVO for further analysis. Starting with a mutation radius of 15–30 % of the parameter range may help during the initial studies, indicating which values to use during the following analysis. We may reduce the mutation radius as soon as we feel that we are close to an interesting region in the parameter space.

5.2 Metal Forming

Iryna Kmitina and Tatiana Popova

With increasing demands placed on metal forming companies, and the growing complexity and variety of their products, simulation of forming processes is an increasingly important field. Understanding how loads will act on a part is very important for sizing forming tools and determining the process borders. Simulations are used to control the quality of the final product at an early stage of the process development, and their flexibility enables quick changes of process parameters, and the evaluation of their effects. Here Bionic Optimization in combination with metal forming simulation can help metal factories avoid defects in their production lines, reduce testing and expensive mistakes, and improve efficiency in the metal forming process.

An example of the metal forming simulation is represented in Fig. 5.7. This example shows a hot forging process. This simulation was made using a finite-element-method (FEM) solver and a finite-volume-method (FVM) solver. Hot forging occurs at a temperature above the recrystallization point of the metal. For mass production, fully automated multi-stage presses are unmatched worldwide.

This example was created and run with Simufact.forming software: a customized software solution for the analysis and optimization of forming processes. Forming processes have a high potential for optimization. This section presents possibilities for the verification of forming processes optimizations.

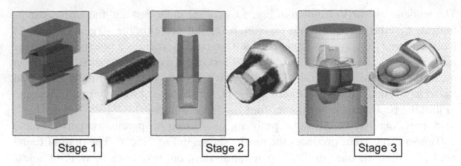

Stage 1 Stage 2 Stage 3

Fig. 5.7 Multi-stage forming process—hot forging (Schmiedag GmbH Hagen)

5.2.1 Deep Drawing

Deep drawing is a method of metal sheet forming. In this process a sheet metal blank is radially drawn into a hollow cup (can) with a forming die and the axial mechanical action of a punch. The end form is achieved by redrawing the intermediate form through a series of dies (Ping et al. 2012).

In our example the workpiece is a blank with predefined dimensions of radius and height. The deep drawing process contains many components and steps. The first forming, shown in Fig. 5.8, uses the following tools: a die (ring), a blank holder, and two punches, which move together during the first stage. The blank lies between the die and holder and then is drawn into a forming die. In the next stage the tools include: a forming die with a smaller diameter, the first punch as a blank holder, and a moving second punch. This intermediate form goes through three ring-dies that make the can thinner. The second punch and bottom-die are used on the bottom forming of a can, the last stage, after stretching.

Here, the optimization task is to achieve a uniform wall thickness distribution at the can after the first forming stage, dependent on tool friction. The workflow of the outer loop optimization process (Sect. 4.1.2), including the optimization method and the Simufact simulation tool, is shown in Fig. 5.9.

Workflow description:

- The workflow starts with a run of control program (optimizer). For example, when using PSO, in the first step, the values of input parameters will be selected randomly within a certain range.
- The changed values are rewritten in the Simufact input data file.
- The simulation job is run in batch mode.
- After a simulation job is finished, the optimizer receives the output file of the last simulation increment, converts it, and reads the results. If these results do not satisfy the restrictions, then the goal value will be recalculated—maximized for a minimization task or minimized for a maximization task (cf. Sect. 2.9). In this way the unsuitable set of input parameters will be restricted, and next cycle of optimization process will be executed. If activated, the program will verify the

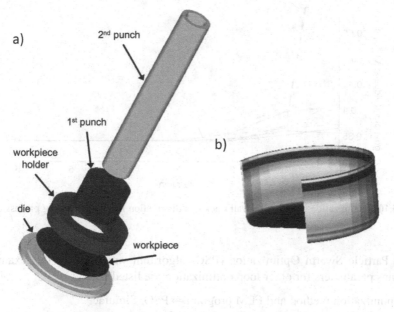

Fig. 5.8 The first stage of the deep drawing process. (**a**) Tools of the first step of deep drawing. (**b**) First intermediate form of the workpiece—simulation indicates the effective plastic strain

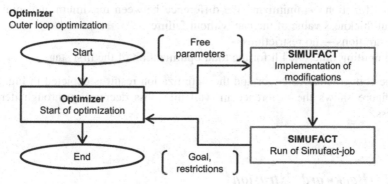

Fig. 5.9 Optimization workflow for can optimization with Simufact

completed job status by checking a stop criterion, for example, if the tolerance between the new and the old fitness values has been reached, the optimizer will be stopped. Otherwise, the input parameters will be recalculated and next cycle of optimization process will run.

Significant variables that can be used for optimization of the deep drawing process include: the properties of sheet metal, blank holder force, tool friction, punch speed, the blank diameter to punch diameter ratio, the sheet thickness, the clearance between the punch and the die, and the punch and die corner radii.

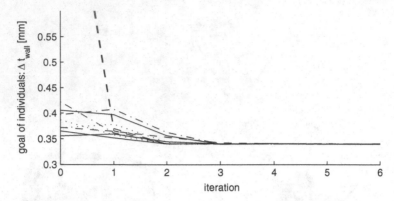

Fig. 5.10 Optimization of the can's wall thickness distribution in a deep drawing process using PSO

A Particle Swarm Optimization (PSO) algorithm was used for this example. Settings parameters for outer loop optimization are listed below:

- Optimization method and FEM program → PSO, Simufact
- Optimization parameters → 10 particles, 6 iterations
- Input parameters and parameter range → friction coefficients of first punch and die [0.05...0.4]
- Goal function → minimizing the difference between maximum and minimum wall thickness value of the can without failure
- Restrictions → no restrictions
- Computing time → 14 h for the total optimization of the first stage

The convergence behavior and the optimization result is depicted in Fig. 5.10. The figure shows the difference in wall thickness decreasing through iterative process.

5.2.2 Backward Extrusion

Backward extrusion is a widely used cold forming process for the manufacturing of hollow symmetrical, cylindrical products. It is usually performed on high-speed and accurate mechanical presses. The punch descends at a high speed and strikes the workpiece, extruding it upwards by means of high pressure. The die ring helps to form the tube wall. The thickness of the extruded tubular section is a function of the clearance between the punch and the die (Barisic et al. 2005). A schematic outline of backward extrusion process is presented in Fig. 5.11.

For the simulation of the backward extrusion process, a simplified process model could be used. For instance, all the punch parts could be represented as one single part. All the tools of the backward extrusion process could be divided into three

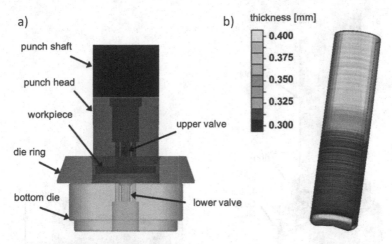

Fig. 5.11 The backward extrusion process. (**a**) Tools of backward extrusion. (**b**) Can—result of a backward extrusion simulation

groups according to their functions: punch, ring and housing base. The workpiece is represented by an aluminum blank. Figure 5.12 illustrates the components of the backward extrusion process in the commercial code Simufact.

The output of the process is the can with its wall thickness. The wall thickness distribution depends on the tool dimensions. Parameter variations cause a thickness distribution. Figure 5.13 depicts all the dimensions of the tools that could be variables for the backward extrusion process optimization. In addition, the thickness of the workpiece may be modified as well.

The goal of the optimization is to minimize the mass, using a PSO study. As a first restriction the required can length of $l = 201$ mm should be reached in the backward extrusion process. Furthermore, the final can has to resist an inside pressure of $p_2 = 21.6$ bar, without large deformations up to $p_1 = 18$ bar. All restrictions are handled by a penalty method (Sect. 2.9). Figure 5.14 shows the fitness value convergence through iterations of Bionic Optimization strategy—PSO (16 iterations × 18 particles = 288 simulation runs). In this example, the fitness value represents the workpiece mass. Optimization is obtained through modifications to the geometry. The figure contains the worst, average and best particles curves. The Bionic Optimization method PSO finally proposes an 21 % mass reduction from the initial 38 g to the optimized mass of 30.0 g.

Forming processes simulations require long calculation times because of the complexity of these problems. Methods to reduce the calculation time lose quality in the final result with simplifications of their models. The length of one calculation can restrict optimization possibilities. Consequently, for increasing the use of forming processes optimization, simulation acceleration is needed. On the other hand, adaption of optimization strategies to specific problems or hybrid optimization procedures can help to reduce the number of jobs in an optimization. In Sect.

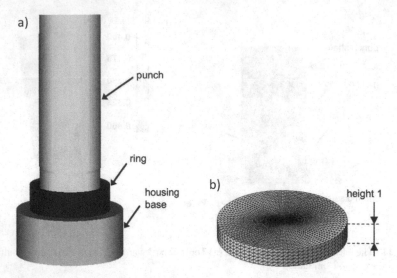

Fig. 5.12 Simulation of the backward extrusion with a simplified model. (**a**) Tools of backward extrusion. (**b**) Meshed workpiece

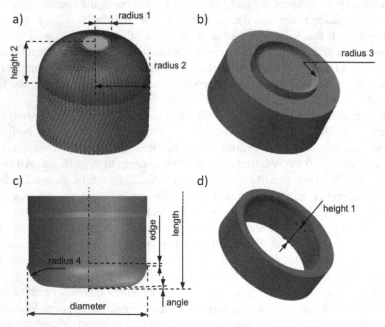

Fig. 5.13 Free parameters at backward extrusion tools to achieve an optimized shape of the can. (**a**) Bottom forming tool. (**b**) Lower die. (**c**) Punch. (**d**) Ring

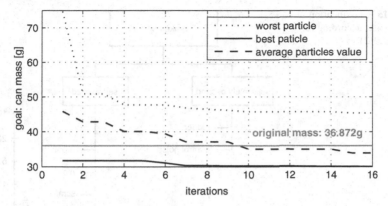

Fig. 5.14 Minimization of the can's mass in a backward extrusion process with PSO

2.7, Example 2.3, we show the advantages of combining PSO and the meta-modeling technique for optimization of the backward extrusion process.

Forming processes have a high potential for optimization. This section demonstrated two examples of the forming processes optimization, which present promising results. Further developments of this kind of process optimization are possible in the future with simulation acceleration.

5.3   Brake Squeal

Simon Gekeler, Oskar Glück, and Ashish Srivastava

When designing brake systems in automotive industry, the biggest challenge, after the brake's performance, is comfort. That is why brake noise, especially brake squeal, is one of several problems engineers face in development. The main cause of brake squeal in disc brakes is the sliding contact between brake pads and the disc, which causes an imbalance of forces, and results in amplified oscillation. Theoretically, higher damping and a less stiff system could reduce the tendency of a brake to squeal. Since this directly affects the efficiency of the brake and the performance of the car, it is necessary to find an optimal configuration of the main parameters, providing high brake performance with less squeal. Here Bionic Optimization strategies help to provide solutions (Losch 2013; Thelen 2013).

5.3.1   Types of Brake Noise

Figure 5.15 shows a classification of the brake noises by their cause and the relevant frequency range, according to (Zeller 2012):

Fig. 5.15 Classification of brake noise, according to Zeller (2012)

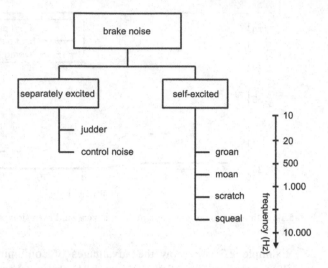

- Judder: Low frequency oscillations up to 500 Hz. It occurs during slight to medium deceleration from high velocities as vibration in the steering wheel or the brake pedal is accompanied by a humming sound.
- Control noise: Operating noise of the ABS-system, which is acoustically uncritical. A response in dangerous situations is required but may not distract the driver.
- Groan: Caused by the alternating stick and slip of the brake pads on the disc (stick-slip-effect).
- Squeal: Occurs at frequencies higher than 1000 Hz. It is almost not overlaid by any other road noises due to its appearance at low driving speeds and thus is particularly bothersome. The cause stems from self-excited vibrations based on instabilities in the friction characteristic of the brake system. These vibrations appear at critical interactions between in-plane-modes and by resonance of bordering components caused out-of-plane-modes (Zeller 2012).

Judder and control noise results from wavering brake torque. They are a combination of vibration and accompanying sounds and occur in lower frequency ranges. Groan, moan, scratch and squeal are self-excited brake noises in the acoustical area, which are caused by dynamical instabilities of the brake system. They occur mainly at low driving speed, e.g., slowing down at traffic lights (Schlagner 2010).

5.3.2 Modeling of Brake Squeal

The setup of a simulation model and the emergence of brake squeal are explained by reference to an example of a brake disc model. With the help of complex Eigen

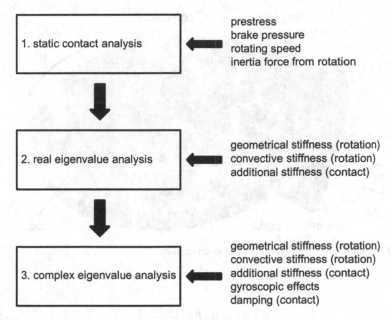

Fig. 5.16 Procedure of simulation of brake squeal (Carvajal et al. 2014)

value analysis, all unstable Eigen frequencies of a system can be encountered during one simulation run. The results serve as indication of the instability of the system and reveal a possible squeal tendency, which in reality does not necessarily have to occur. Nevertheless, experiments have shown a good correlation between the results of the complex Eigen value analysis and real-world squeal noises. Figure 5.16 shows a basic procedure of the simulation of brake squeal, e.g., with PERMAS, a Finite Elements analysis system developed by INTES GmbH (https://www.intes.de/, April 2015).

To explain the analysis of brake squeal a simplified brake model is used (Fig. 5.17). The model consists of a rotating disc with constant angular frequency ω_{rot}. The disc is constrained in all spatial directions at its inner diameter. The pads are guided in x- and z-direction (in-plane) by rigid elements fixed to the caliper. Normal to the disc (out-of-plane) they are supported against the caliper by springs. For the contact between disc and pads, a constant sliding friction μ is assumed. The pads are loaded by a constant pressure p (INTES GmbH 2014).

1. To determine the contact state and contact forces during the first computation step, a static analysis with contact and friction under brake pressure and a rotation ω_{rot} is performed. The relative velocity v_{rel} between brake pad v_{pad} and brake disc v_{disc} is calculated and defines the direction of the friction forces F_f. We get the magnitude of these forces according to Coulomb's law of friction in Eq. (5.7). Here the brake pressure is acting as normal force F_N.

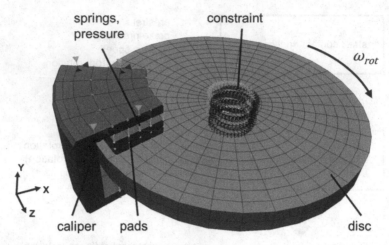

Fig. 5.17 Simplified brake disc model and its constraints

$$F_f = \mu \cdot F_N \cdot \frac{v_{rel}}{|v_{rel}|} \tag{5.7}$$

The contact state is frozen for succeeding computation steps and provides additional amounts to the stiffness and damping matrices.

In Fig. 5.18 the forces on the disc model and the relative velocity between the disc and the pads are displayed. Here the stiffness term k and damping term c represent the properties of the caliper.

As result of the rotating disc, we get the convective stiffness $\mathbf{K}_{conv}$ and also the damping $\widetilde{\mathbf{C}}_{gyro}$ and stiffness $\mathbf{K}_{gyro}$ caused by gyroscopic effects in a second computation step. Those additional terms will be considered in the following Eigen value analyses.

2. The Eigen modes of the brake disc can be determined by a real Eigen value analysis. The differential equation for undamped free vibrations in Eq. (5.8) together with the assumption of a periodic response in Eq. (5.9) deliver the Eigen value problem in Eq. (5.10), which is extended by the previous calculated stiffness terms resulting in Eq. (5.11) (Nasdela 2012).

$$\mathbf{M\ddot{u}} + \mathbf{Ku} = 0 \tag{5.8}$$

$$\mathbf{u} = \boldsymbol{\varphi} \exp(i\omega t) \tag{5.9}$$

$$(-\lambda_i \mathbf{M} + \mathbf{K}) \cdot \boldsymbol{\varphi}_i = 0 \ with \ \lambda_i = \omega_i{}^2 \tag{5.10}$$

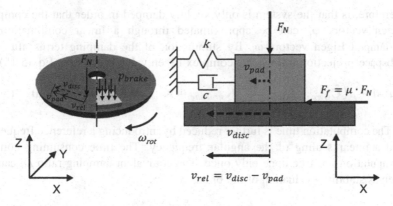

Fig. 5.18 Representation of forces on the brake (Losch 2013)

$$\left(-\lambda \mathbf{M} + \left(\mathbf{K}_{structure} + \mathbf{K}_{conv} + \mathbf{K}_{Gyro}\right)\right) \cdot \mathbf{\Phi} = 0 \; with \; \mathbf{\Phi} = \left(\boldsymbol{\varphi}_1 \ldots \boldsymbol{\varphi}_n\right) \quad (5.11)$$

We look at the non-trivial-solutions:

$i = 1 \ldots n_{eig}$	with n_{eig}, the number of interesting Eigen values
λ_i	Eigen values
$\omega_i = \sqrt{\lambda_i}$	Eigen angular frequency
$\boldsymbol{\varphi}_i$	Eigen vector

3. From the differential equation for damped free vibrations in Eq. (5.12) and the approach in Eq. (5.13), we derive the complex Eigen value problem in Eq. (5.14).

$$\mathbf{M}\ddot{\mathbf{u}} + \mathbf{C}\dot{\mathbf{u}} + \mathbf{K}\mathbf{u} = 0 \qquad (5.12)$$

$$\mathbf{u} = \boldsymbol{\varphi}^* \exp(\mu t) \qquad (5.13)$$

$$\left(\mu^2 \mathbf{M} + \mu \mathbf{C} + \mathbf{K}\right) \cdot \mathbf{\Phi}^* = 0 \; with \; \mathbf{\Phi}^* = \left(\boldsymbol{\varphi}_1^* \ldots \boldsymbol{\varphi}_n^*\right) \qquad (5.14)$$

Non-trivial-solutions are:

$\mu_{1/2,i} = -\delta_i \pm i\omega_{d,i}$	complex Eigen value
δ_i	damping coefficient
$\omega_{d,i}$	angular frequency of the damped oscillation
$\boldsymbol{\varphi}_i^*$	complex Eigen vector

To reduce the amount of degrees of freedom, and thus the computation time, the complex Eigen value problem in Eq. (5.14) is performed in a subspace according to the Craig-Bampton-Method (Nasdela 2012). A prerequisite, therefore, is that the system is only slightly damped in order that the complex Eigen vectors $\boldsymbol{\varphi}_i^*$ can be approximated through a linear combination of undamped Eigen vectors $\boldsymbol{\varphi}_i$. By summation of the damping terms after the subspace-projection follows the complex Eigen value problem in Eq. (5.15).

$$\left(\mu^2 \tilde{\mathbf{M}} + \mu\left(\tilde{\mathbf{C}}_{Gyro} + \tilde{\mathbf{C}}_{\mathrm{mod}}\right) + \left(\tilde{\mathbf{K}}_{structure} + \tilde{\mathbf{K}}_{conv} + \tilde{\mathbf{K}}_{Gyro}\right)\right) \cdot \boldsymbol{\Phi}^* = 0 \quad (5.15)$$

The computation time is further reduced by introducing a reference frequency and a linear scaling of the angular frequency. The time consuming contact computation must be done only once. The equivalent damping ratio D_i can be seen as stability evaluation of the brake system.

$$D_i = \frac{\delta_i}{|\omega_{d,i}|} \quad (5.16)$$

The equivalent damping ratios D_i plotted over the rotational speed results in separate Campbell-Diagrams for each mode. The following stability conditions, depicted in Fig. 5.19, can occur:

– The equivalent damping ratio of one mode is positive for the whole range of rotational speed:
  $D > 0$, mode is stable, no tendency to cause squealing (Fig. 5.19, mode 1)
– At least one equivalent damping value is negative:
  $D < 0$, mode is unstable, tendency to cause squealing (Fig. 5.19, mode 2 and mode 3). However some modes can be ignored as they do not excite squealing.

Thus the real part of the Eigen value μ_i provides information of the stability, and the imaginary part describes the angular frequency of this mode.

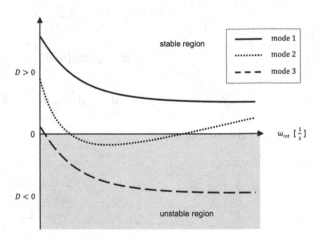

Fig. 5.19 Campbell diagram with equivalent damping ratio of Eigen modes depending on rotational speed

5.3.3 Minimizing the Risk of Brake Squeal Using Bionic Optimization

The goal of the optimization is to provide a design with the least possibility to cause squealing, without affecting brake performance. Negative equivalent damping ratios of Eigen modes need to be avoided. Thus we want to find a design where all relevant modes have positive equivalent damping ratios. Furthermore it is necessary to understand the importance of how negative a damping characteristic might be. Even small negative damping amounts could cause unstable vibration behavior. With respect to the mentioned problems, it is vital to frame a suitable goal function which meets these requirements.

We assume an existing design of a brake system. Hence, as a second goal, we do not want to have large modifications in the design. Alterations to the free parameters should be as small as possible, while minimizing the squealing tendency. To avoid the complexity of a Multi-Objective Optimization, we transform our first goal (reduce squealing) into a restriction handled by the penalty method (Sect. 2.9). This means, while achieving our goal of a minimum change in the current design, we need to make sure that all Eigen modes have a positive (above $\varepsilon = 10^{-4}$) equivalent damping ratio in a range of a rotational speed ω_{rot} from 0.5 Hz to 5.0 Hz. Otherwise, a penalty value is applied on our goal (Losch 2013). For our simplified brake disc model, we define Young's modulus of the brake pads material E and the friction coefficient μ between the pads and the disc as free parameters. Now we can describe our optimization problem as

$$min_{\Delta\mu,\Delta E}\left(\Delta\mu^2 + \Delta E^2\right) \tag{5.17}$$

$$D(\omega_{rot}, Mode) \geq \varepsilon \tag{5.18}$$

$$\omega_{rot} = 0,5\ldots 5Hz, Mode = \{1, 2, \ldots, 23\},$$

where $\Delta\mu$ and ΔE are the normalized changes of the free parameters in their defined limits $[\mu_{min}, \mu_{max}]$ and $[E_{min}, E_{max}]$ according to

$$\Delta\mu = \frac{\mu - \mu_0}{\mu_{max} - \mu_{min}} \tag{5.19}$$

$$\Delta E = \frac{E - E_0}{E_{max} - E_{min}} \tag{5.20}$$

The Campbell diagram in Fig. 5.20 indicates the squealing tendency of our simplified brake disc example. Following the numerical results, the complex modes 19 and 21 have negative equivalent damping ratios. The complex mode 11 and mode 21 have low damping values, too. They risk becoming unstable when varying parameter values in an optimization process. The corresponding mode shapes are depicted in Fig. 5.21.

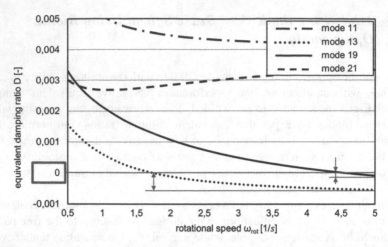

Fig. 5.20 Campbell diagram with critical complex Eigen modes for the simple brake disc model (Losch 2013)

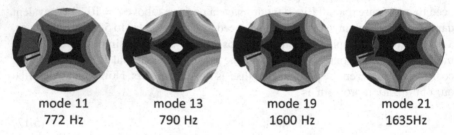

mode 11	mode 13	mode 19	mode 21
772 Hz	790 Hz	1600 Hz	1635Hz

Fig. 5.21 Mode shapes of the critical complex Eigen modes (Losch 2013)

Table 5.1 Parameter values of original and optimized brake design (Losch 2013)

	Friction coefficient μ (–)	Young's modulus pad E (MPa)	Change in free parameters $(\Delta\mu^2 + \Delta E^2)$
Original design	0.700	500.0	0
Optimized design	0.476	481.1	1.411×10^{-1}

A Particle Swarm Optimization with 20 particles and 20 iterations gives us a proposal of a non-squealing brake design, with few changes compared to the original. Table 5.1 shows the changes in parameter values. The Campbell diagram (Fig. 5.22) shows that no negative equivalent damping ratio occurs within the range of rotational speed.

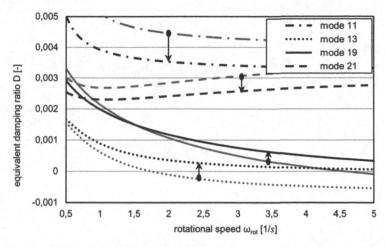

Fig. 5.22 Campbell diagram with critical complex Eigen modes of original and optimized brake design (Losch 2013)

References

Barisic, B., Car, Z., & Cukor, G. (2005). Analytical, numerical and experimental modeling and simulation of backward extrusion force on Al Mg Si 1. In K. Kuzman (Ed.), *5th International Conference on Industrial Tools, Velenje* (pp. 341–347).

Berkeley. (2011). *Kobe earthquake 1995/01/16 20:46.* Station Takarazuka: Earthquake Engineering Research Center (PEER). http://ngawest2.berkeley.edu/

Carvajal, S., Wallner, D., Wagner, N., Klein, M. (2014). Hervorragender Geräuschkomfort durch Simulation – Neue Methoden zur Berechnung von Stabilitätskarten. In INTES GmbH, & Dr. Ing. h.c. F. Porsche AG (Eds.). Bamberg: NAFEMS Konferenz.

INTES GmbH. (2014). *PERMAS examples manual.* INTES Publication No. 550, Stuttgart. PERMAS Version 15.00.112.

Losch, E. (2013). *Inner Loop-Optimierung einer Bremsscheibe mit kontrollierter Geräuschentwicklung.* Master thesis, Reutlingen University.

Nasdela, L. (2012). *FEM-Formelsammlung Statik und Dynamik: Hintergrundinformationen, Tipps und Tricks.* Wiesbaden: Vieweg+Teubner Verlag.

Ping, H., Ning, M., Li-zhong, L., & Yi-guo, Z. (2012). *Theories, methods and numerical technology of sheet metal cold and hot forming: Analysis, simulation and engineering applications.* London: Springer.

Schlagner, S. (2010). *Schnelle Charakterisierung des Geräuschverhaltens von KFZ-Scheibenbremsen.* Aachen: Shaker.

Schmiedag GmbH Hagen. *picture from: Schmiedag GmbH Hagen, simufact.demos 2013, hot forging – multistage processes.*

Steinbuch, R. (2011). Bionic optimisation of the earthquake resistance of high buildings by tuned mass dampers. *Journal of Bionic Engineering, 8,* 335–344.

Thelen, M. (2013). *Outer Loop-Optimierung einer Bremsscheibe mit kontrollierter Geräuschentwicklung.* Master thesis, Reutlingen University.

Zeller, P. (2012). *Handbuch Fahrzeugakustik: Grundlagen, Auslegung, Versuch.* Wiesbaden: Vieweg+Teubner.

Chapter 6
Current Fields of Interest

Rolf Steinbuch, Iryna Kmitina, and Nico Esslinger

Research in structural optimization started with the onset of structural mechanics. The availability of powerful computers and efficient simulation tools such as FEM, BEM or MBS have led to many optimization methods using closed loops that require no or little user interaction during their implementation. Today, much of the commercial and open source software has more or less integrated the optimization modules, based on various principles. As we can see in many other fields, this possibility did not only result in positive response from the users, but it also increased the demand for easily enabled optimization. We observe an increasing demand for optimization and as a consequence a growing number of tools that help designers to improve on their original ideas.

Gradient based optimization and sensitivity studies are common options in CAE-systems today, and so we recommend that engineers change their software packages, if their existing system does not provide these capabilities.

External software packages such as optiSLang (http://www.dynardo.de/soft ware/optislang, retrieved 15.04.2015) or the codes developed by Reutlingen Research Institute (RRI) or other institutes provide enhanced possibilities supplied by the different codes by outer loop optimization (cf. Sect. 4.1.2). We believe that such overlays will contribute fundamentally to the expansion of optimization in structural design and other engineering fields as well.

If we review the research done in the field of optimization, the following topics appear to be the focus of current development:

– Optimization under uncertainties, taking into account the inevitable scatter of parts, external effects and internal properties. Reliability and robustness both

R. Steinbuch (✉) • I. Kmitina • N. Esslinger
Hochschule Reutlingen, Reutlingen Research Institute, Alteburgstraße 150, 72762 Reutlingen, Germany
e-mail: Rolf.Steinbuch@Reutlingen-University.DE; Iryna.Kmitina@Reutlingen-University. DE; Nico.Esslinger@Reutlingen-University.DE

© Springer-Verlag Berlin Heidelberg 2016
R. Steinbuch, S. Gekeler (eds.), *Bionic Optimization in Structural Design*,
DOI 10.1007/978-3-662-46596-7_6

have to be taken into account when running optimizations, so the name Robust
Design Optimization (RDO) came into use.

- Multi-Objective Optimization (MOO) handles situations in which different
  participants in the development process are developing in different directions.
  Typically we think of commercial and engineering aspects, but other constella-
  tions have to be looked at as well, such as comfort and performance or price and
  consumption.
- Process development of the entire design process, including optimization from
  early stages, might help avoid inefficient efforts. Here the management of virtual
  development has to be re-designed to fit into a coherent scheme.
- Further improvement of the bionic and other related non-deterministic strategies, especially the reduction of the number of jobs and increasing quality of the
  prediction, will undergo continual evolution.

There are many other fields where interesting progress is being made. We limit
our discussion to the first three questions, as we discuss the performance of bionic
methods throughout this book, especially in Sect. 3.1.

6.1 Reliability and Robustness

Iryna Kmitina

Uncertainty is inevitable in engineering design. Every component, every material
and all load sets are not given by exact data, but tend to scatter around some
predefined values. Therefore research about design under uncertainty has been
growing over the last years and is now used in a wide range of fields from simple
product components to designing complex systems. Terms such as "Robust
Design" and "Reliability Based Design Optimization" have been introduced in
some design software packages. But their application to parametric uncertainty is
difficult and limited. Robust design is mainly exploited to improve the quality of a
product and to achieve the required level of performance. While this can be done by
minimizing the effect of the scatter; however, the causes are not eliminated. The
reliability-based design tries to keep the failure probability below an acceptable
level.

We have learned already that numerical optimization of mechanical designs
using simulation systems such as FEM requires much computing power in terms of
jobs, capacity and time. The additional effort to provide sufficient information for
the evaluation of the reliability or robustness of the design may become even larger.
In consequence, efficient strategies must be used to ensure reliability or robustness.

6.1.1   Reliability-Based Design

Reliability-based Design Optimization (RBDO), as one paradigm of design under uncertainty, seeks optimal designs with low probabilities of failure within the expected scatter of the produced parts. Mathematically, a basic formulation of RBDO is described as (Wang et al. 2010; Du and Chen 2004):

$$\min_{d,\mu_X} f(\mathbf{d}, \mathbf{X}, \mathbf{C}) \tag{6.1}$$

$$\text{subject to Prob}\{g_i(\mathbf{d}, \mathbf{X}, \mathbf{C}) \geq 0\} \geq R, i = 1, 2, \ldots, N_g$$

$$\mathbf{d}^L \leq \mathbf{d} \leq \mathbf{d}^U, \quad \mu_X^L \leq \mu_X \leq \mu_X^U,$$

where

- $f(\cdot)$ is an objective function;
- $\mathbf{d}$ is a vector of deterministic design variables;
- $\mathbf{X}$ is a vector of random design variables;
- $\mathbf{C}$ is a vector of random parameters (not changeable and not controllable in the design process);
- μ_X is the vector of mean values of random design variables;
- g_i is the ith limit state function and N_g is the total number of limit state functions;
- Prob$\{\cdot\}$ denotes a probability of failure;
- R is desired reliability level.

As we know that a reliability analysis is computationally expensive, we need to find relatively efficient methods to handle it. Among such methods, analytical approximations of the goal and the restrictions are often used. The limit state function, for example, is represented by a first or second order Taylor series expansion, so we speak about First Order Reliability Method (FORM) or Second Order Reliability Method (SORM). It is often assumed that the higher order estimation produces precise estimations. Unfortunately this is not always true.

The approximation methods consist of just a few steps. In the first step, the random variables are transformed from their original distribution into a standard normal distribution by means of the so-called Rosenblatt transformation. This corresponds to the replacement of the original distribution with a normal distribution with the same mean and standard deviation, then mapping this new random variable to a normalized one. Now all random variables cover the same range, disregarding their real physical values (Fig. 6.1a). The resulting multidimensional distribution is sketched in Fig. 6.1b. All random variables cover the same range. There is no difference between their appearances. In addition we now use FORM or SORM to quantify the measure of the failure area by approximating the restriction by linear or quadratic hyper-surfaces shown in Fig. 6.1b as well (Gekeler and Steinbuch 2014).

The shortest distance from the constraint function $g(p_1, p_2) = 0$ to the origin in a standard normal space is called reliability index β. The point that has the highest probability density on the constraint function is called the Most Probable Point

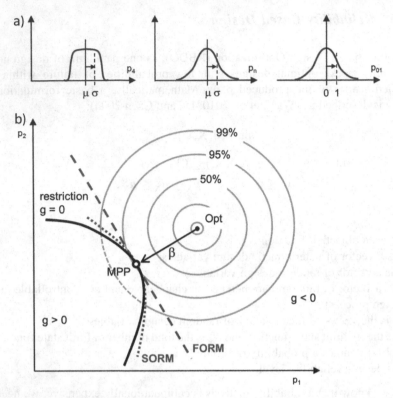

Fig. 6.1 Transformation of random variables to a normalized multidimensional distribution. (**a**) Rosenblatt transformation of random variables. (**b**) Optimum (Opt), Restriction, MPP, FORM and SORM

(MPP). A design can fall into the safe region that is defined by $g(p_1, p_2) < 0$—reliability, or into the forbidden region $g(p_1, p_2) > 0$—failure.

We should realize that the use of FORM or SORM is not necessarily conservative. In Fig. 6.3 as we indicate, there are regions in the 2D space which are not defined as violating the given restriction $g > 0$ by FORM or SORM.

6.1.2 Robust Design

Robust Design Optimization (RDO) seeks a product design which is not too sensitive to changes of environmental conditions or noise. The task of robust design is different from reliability-based design. RDO tries to minimize the mean and the variation of the objective function simultaneously under the condition that constraints are satisfied (Wang et al. 2010; Tu et al. 1999). Mathematically a basic formulation of RDO is described as

$$\min_{d,\mu_X} f\left(\mu_f(\mathbf{d},\mathbf{X},\mathbf{C}),\sigma_f(\mathbf{d},\mathbf{X},\mathbf{C})\right) \tag{6.2}$$

$$\text{subject to } g_i(\mathbf{d},\mathbf{X},\mathbf{C}) \le 0, i = 1,2,\dots,N$$

$$\mathbf{d}^L \le \mathbf{d} \le \mathbf{d}^U, \quad \mu_X^L \le \mu_X \le \mu_X^U,$$

where μ_f is the mean value and σ_f is standard deviation of the objective function, N is the number of deterministic constraints. This is a Multi-Objective Optimization (MOO, cf. Sect. 6.2) problem. We often manage it with the weighted sum method or another appropriate method (Du et al. 2004).

6.1.3   Reliability and Robustness Integration

For optimization under uncertainty, it is necessary to take both the probabilistic design constraints and the design objective robustness into account. In Fig. 6.2 one can observe that unreliable parts are not robust, as they fail to comply with the restrictions. This corresponds to unacceptable values of the objective (Gekeler and Steinbuch 2014).

The integration of both robustness and reliability considerations can be expressed using Eqs. (6.1) and (6.2)

$$\min_{d,\mu_X} f\left(\mu_f(\mathbf{d},\mathbf{X},\mathbf{C}),\sigma_f(\mathbf{d},\mathbf{X},\mathbf{C})\right) \tag{6.3}$$

$$\text{subject to Prob}\{g_i(\mathbf{d},\mathbf{X},\mathbf{C}) \ge 0\} \ge R, i = 1,2,\dots,N_g$$

$$\mathbf{d}^L \le \mathbf{d} \le \mathbf{d}^U, \quad \mu_X^L \le \mu_X \le \mu_X^U.$$

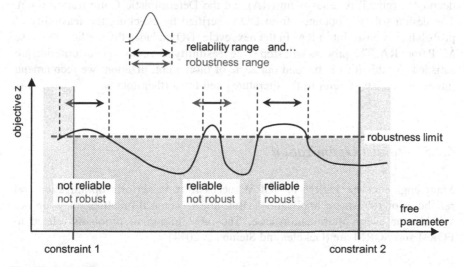

Fig. 6.2 Reliability and robustness

In order to overcome the difficulty of choosing the weighting factors, a unified framework method has been suggested (Wang et al. 2010).

6.1.4   A Sketch of a Formulation of a Unified Reliability and Robustness Strategy

To overcome the difficulty of choosing weighting factors, Wang et al. (2010) tried to formulate general unified framework for integrating reliability-based and robust design. The optimization task is to minimize the probabilistic objective function under the condition that constraints are satisfied, in other words, the design points appear in the safe region. In the case with normal distributed objective functions a unified framework is provided by

$$\min_{d, \mu_X} \quad \mu_f + k * \sigma_f$$

$$\text{subject to Prob}\{g_i(\mathbf{d}, \mathbf{X}, \mathbf{C}) \geq 0\} \geq R, i = 1, 2, \ldots, N_g$$

$$\mathbf{d}^L \leq \mathbf{d} \leq \mathbf{d}^U, \mu_X^L \leq \mu_X \leq \mu_X^U, \tag{6.4}$$

where k is a constant expressing the weighting of the mean and standard deviation. This weight also predicts the satisfaction's probability of objective function. For instance, $k = 3$ means that the $\text{Prob}\{f \leq \mu_f + k\sigma_f\} = 99.87\ \%$.

The Sequential Optimization and Reliability Assessment (SORA) method may be used to solve the optimization problem with normal distributed objective functions (Yin and Chen 2006). The SORA approach consists of an idea called decoupled reliability assessment (RA) and the Deterministic Optimization (DO). The design solution obtained from DO is verified by checking the feasibility of probabilistic constraint in RA. In the next cycle, DO includes the predicted inverse MPP from RA. The process will stop if the feasibility and convergence criterion are satisfied. As this idea is beyond our topic of bionic optimization, we recommend interested readers to refer to the literature cited for further details.

6.1.5   Robust Optimization

Many engineers use FORM or SORM successfully to perform optimization and reliability or robustness applications. But, due to some difficulties, they are not suitable for every optimization case. The most important problems related to FORM and SORM are (Gekeler and Steinbuch 2014):

Fig. 6.3 Second restriction and non-conservativeness of FORM (F) and SORM (S)

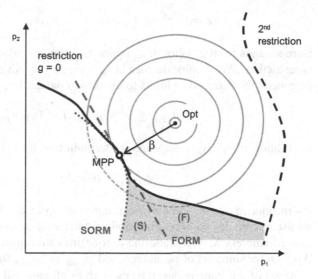

- scattering input data have to be independent when they are considered as random variables. They must follow a normal distribution or have been transformed into a normal distribution;
- the linear or quadratic approximation of the restrictions hyper-plane may not be conservative. In Fig. 6.3 (F) indicates the region where FORM and SORM are not conservative, while (S) adds the region where SORM is not conservative;
- the normalization of the random variables requires a good guess of the mean and standard deviation of the multidimensional random variables which may be found only after a large number of tests;
- the approaches primarily hold only for one critical restriction, and they may fail or become less applicable as soon as there is a second restriction active as shown in Fig. 6.3.

As the proposed approaches to carrying out reliability and robustness studies consume much time and computing power, faster steps to come up with acceptable results were proposed (Gekeler and Steinbuch 2014). These proposals, found by the optimization techniques, may be used as input for manufacturing without having to consider uncertainty at all.

To take into account stochastic problems, a more general definition was suggested. The objective function is described as:

$$\mathbf{z} = \mathbf{z}(p_1, p_2, \ldots, p_{np})^T, \tag{6.5}$$

where $\mathbf{z}$ is a vector composed of two other vectors, $\mathbf{s}$ and $\mathbf{r}$:

$$\mathbf{z} = (\mathbf{s}, \mathbf{r})^T = \left(s_1, s_2, \ldots, s_{n_g}, r_1, r_2, \ldots, r_m \right)^T, \tag{6.6}$$

Here $\mathbf{s}$ stands for the vector of n_g optimization goals, while $\mathbf{r}$ represents the set of m restrictions. We confine the idea here to single-objective optimization i.e. $\mathbf{s} = s_1$. In general there are given limits to the design parameters like before

$$p_{i,\min} \leq p_i \leq p_{i,\max}, i = 1 \ldots n_p. \tag{6.7}$$

In addition all p_i may show some scatter indicated by

$$p_i = p_i \pm \Delta p_i. \tag{6.8}$$

As mentioned above, some authors, e.g., (Wang et al. 2010), distinguish sets of non-scattering design or optimization parameters $\mathbf{d}$, scattering design or optimization parameters $\mathbf{X}$, and scattering non-optimization parameters $\mathbf{C}$. If one allows $\Delta p_i = 0$ for some set of parameters and $p_{i,\min} = p_{i,\max}$ for another set or even the same set of the same parameters, these three classes will be reduced to one set of optimization parameters $\mathbf{p}$ as proposed in Eqs. (6.5)–(6.8). Some of them do not essentially scatter, and some of them are fixed within their tolerances. This allows for a more simple annotation without losing the generality of the idea.

The main concern of stochastic mechanics is to use a sufficient amount of test data to provide acceptable probabilistic measures. One common and efficient way to solve this problem is using a Response Surface (RS, cf. Sect. 2.7) approach in all components of $\mathbf{z} = (\mathbf{s}, \mathbf{r})^T$. It provides an approximation of the distribution and allows for an estimation of the mean and standard deviation of all the components of $\mathbf{z}$. In this formulation, the goal and the restrictions are defined respectively as $\mathbf{s}$ and $\mathbf{r}$.

In most cases, RS are often first or second order degree polynomials (cf. Sect. 2.7) in the optimization parameters. Since frequently better data are not available, one may use them to perform the reliability or the robustness analysis. The main disadvantage of this approach is that a large number of tests are required (i.e. FE-jobs or experimental measurements). The RS is defined by its coefficients:

$$RS(p_1, p_2, \ldots, p_{np}) = a_0 + \sum_i a_i p_i + \sum_i \sum_{k \leq i} a_{ik} p_i p_k. \tag{6.9}$$

The number of coefficients for a second order Response Surface is

$$n_c = n_p + 1 + (n_p + 1)n_p/2,$$

where n_p denotes the number of optimization parameters. Non-design parameters $\mathbf{C}$ as defined at the beginning of this section might be omitted to reduce the number of studies. To find the RS by a least squares method, the number of tests should be about twice the number of coefficients. In consequence one ought to run approximately n_p^2 tests. For large nonlinear studies and some (e.g., $n_p = 10$) optimization parameters, where one job may take some hours, the total computation time may become absolutely unacceptable. Reduction of the number of coefficients in Eq. (6.9) by omitting the mixed terms to

$$RS(p_1, p_2, \ldots, p_{np}) = a_0 + \sum_i a_i p_i + \sum_i a_{ii} p_{ii}^2 \qquad (6.10)$$

may sometimes help to accelerate the process, as there are only $2n_p + 1$ unknowns and one has to run about $4n_p$ tests. But this simplification may essentially reduce the quality of the approximation. The response surfaces found by any means may be used to estimate the goal or the reliability as shown in Fig. 6.4. The short vertical lines indicate the test data and their distance to the RS.

To continue, it is assumed that the RS are sufficiently good representations of the distribution of the studies' results. The estimation of the reliability by using the RS can be done afterwards. It would be appropriate to assume the RS to be scattering as well. Their standard deviation might be guessed from the deviation of the difference between the RS and the test data

$$\sigma_{RS}^2 = \frac{1}{n-1} \sum_{i=1}^{n} \left(test(\mathbf{p}_i) - RS(\mathbf{p}_i) \right)^2. \qquad (6.11)$$

Here $\mathbf{p}_i$ represents the vector of all design variables at the test # i including the scattering and non-scattering design variables. It is evident, that the better the Response Surface is able to represent the data, the smaller the estimated standard deviation σ_{RS}^2 becomes.

In many cases the optimum and the MPP (cf. Fig. 6.3) coincide. If the random variables are following normal distributions, one may find the probability at parameter values from the mean and the standard deviation. The reliability close

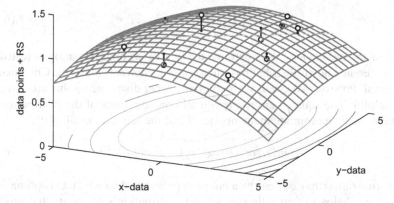

Fig. 6.4 Approximation of a goal or restriction by a second order response surface

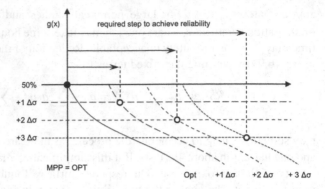

Fig. 6.5 Reliability and scatter, *dots* indicate the combined probability of goal and restriction

to the MPP and Optimum becomes 50 % because $\beta = 0$. In consequence, one has to move away from the MPP along the gradient of the restriction into allowed ($g < 0$) regions. In this way the distance to not only MPP but also to the optimum will be increased to raise the reliability.

A high quality of the Response Surfaces to reproduce the data input makes the prediction of the failure probability more realistic and not over-conservative. In Fig. 6.5, the deterministic reliability curves correspond to 50 % if MPP = OPT. To improve the reliability, the corresponding scatters of the restriction and the proposed design have to be taken into account.

The assumption of a normal distribution of all random variables may cause large standard deviations. This decreases the power of the always doubtful stochastic statements. To reduce both weak components and to improve the performance, better approximations for the distributions could be used. With knowledge of the type of the distributions and their moments, they can be used for some or all optimizations. It is assumed that the random variables, optimization parameters and the scattering design data are each distributed independently. Then the total probability density will be the product of the probability density pr_i of the parameters:

$$Pr_{total} = \prod_i pr_i(p_i), i = 1, \ldots, n_p \qquad (6.12)$$

For all of the parameters' distributions, the quality of approximation is tested by using different known distributions such as, e.g., normal, uniform, Chi-squared, log-normal, Poisson, Maxwell, Weibull or any other distributions that are assumed to be helpful. The squared error between all configurations of the distributions is minimized with appropriate moments (μ, σ) and the test data available:

$$min\left(error((Pr_{total}(p_i) - test(p_i)))^2\right) \qquad (6.13)$$

This minimization may be done by a bionic approach to deal with local optima. This optimization helps to find estimates for the distribution's moments that produce good approximations. Even if this search requires many loops to check all possible combinations of distributions, the time for this search will be smaller compared to the time required for real FE-jobs.

After selecting an appropriate distribution, the problem of estimating the resulting failure probability still remains. For a normal distribution, one can measure the length of the β-vector and compare its length to the standard deviation. For mixed type product distribution and scattering restrictions, a realistic guess of the length and the interpretation of β are required. This question may be solved by some tests of the design on the line between optimum and MPP or along the gradient of the restriction g through the MPP. From these tests, approximations of the distribution and the failure probability will be derived.

In many cases the optima found lie close to a restriction. In these cases, neither reliability nor robustness requirements are fulfilled. If such an optimized design does not provide sufficiently high reliability or robustness, its free parameters must be modified to shift it away from the critical regime. This may be done by translating the parameters along a direction close to the normal β or the gradient of the restriction g from the MPP in Fig. 6.1. The normal on the restriction may not be the direction of the fastest improvement of the reliability as long as the normalized normal distribution is not used. Studies, such as the ones on the response surfaces, may help to give acceptable representations of the preferable position of the design. Care should be taken in the presence of more than one restriction (Fig. 6.3). If other restrictions prohibit feasible solutions near the optimum, we need to search other regions of the parameter space which are large enough to allow for solutions that do not violate any restriction.

Example 6.1 We analyze the bending of an L-Profile fixed at its lower end while a deflection of the upper end of 400 mm is applied. The goal is the minimization of the mass of L-Profile. The length L_1 and thickness T are defined as free parameters (Fig. 6.6).

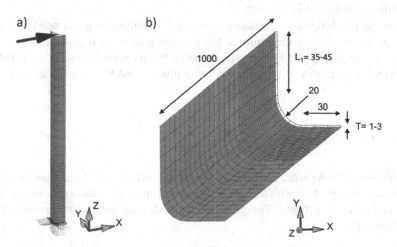

Fig. 6.6 L-Profile under displacement controlled bending load. (a) Overall view. (b) Free parameters L_1 and T

Fig. 6.7 Definition of
constraints on force and
energy

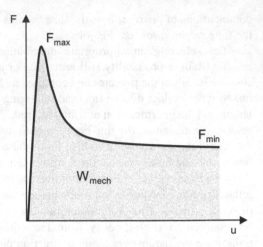

Figure 6.7 indicates the meaning of the constraints on the force and energy.

- Force $F(u) < F_{max}$,
- $F(u_{max}) > F_{min}$
- $W_{mech} > W_{min}$.

In order to generate the corresponding response surfaces, one needs to place variants in the parameter space. This can be done using, for example, the Latin Hypercube Sampling method. Afterward, response surfaces for the goal and the constraints will be generated. Then the restrictions are applied to the Response Surfaces (see Fig. 6.8).

The optimization is done on the response surface of the mass in order to find the deterministic optimum. The search for the optimum without taking into account the scattering is indicated in Fig. 6.9.

Now the reliability and robustness of the optimum must be guaranteed. We do it by stepping away from the limits of the allowed parameter region, following the expected scatter (Fig. 6.10). The quantification of this scatter must be provided by real-world experiences of the manufacturing process and the material quality.

6.1.6   Conclusion

The question of robustness and reliability in optimization problems under uncertainties must be studied with the aim of providing applicable strategies that may be used in the design process. The proposed methods may help to understand of the basic concepts.

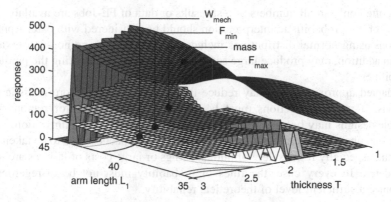

Fig. 6.8 Response Surface with applied restrictions for L-Profile

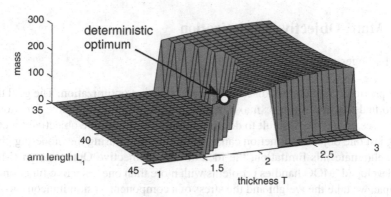

Fig. 6.9 Optimization on RS, which represents the mass in the acceptable parameter region

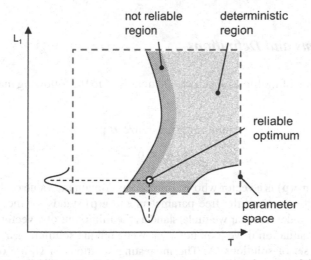

Fig. 6.10 Guess reliability and robustness by the use of the expected scatter of the input data

As often only small numbers of test results or data of FE-Jobs are available, the quality of the probabilistic interpretation should be considered with care. Approximations using normal distributions include the danger of being non-conservative and, in addition, may produce large scatter predictions, thus reducing the predicted reliabilities.

Adapted approximations may reduce the scatter and yield more realistic predictions. If many restrictions must be considered, the search for regions with feasible designs may become more tempting than the original optimization. In all cases, the inherent uncertainties of such stochastic approaches need to be taken into account, especially if the safety of human beings or large costs of failures are to be considered. In every case, the rules of probability must not be disregarded to guarantee a sufficient level of theoretical reliability.

6.2   Multi-Objective Optimization

Nico Esslinger

In the previous chapters, we discussed mono-objective optimization. The goal there was to find the minimum or maximum of a defined scalar objective function. In some cases, it may be difficult to define a problem with just one objective function. Using just one objective function can also lead to a bias during the modeling phase.

To eliminate this limitation, the idea of Multi-Objective Optimization (MOO) was developed. MOO handles problems with more than one objective function. For example, we take the weight and the stress of a component as simultaneous goals in the same optimization study.

6.2.1   Terms and Definitions

The introduction of multiple objective functions lead to the following mathematical problem:

$$\text{minimize/maximize } \mathbf{z}(\mathbf{p})$$

$$\mathbf{g}(\mathbf{p}) \leq 0$$

In this equation $\mathbf{z}(\mathbf{p})$ is a vector whose components contain the value of the different objective functions, $\mathbf{p}$ are the free parameters and $\mathbf{g}(\mathbf{p})$ stands for the constraints. We now have to define what we understand as the minimum of a vector. We avoid this undefined situation if we look not at only one unique solution of the optimization, but at a set of solutions Ω_t. The interesting solutions of Ω_t are often called Pareto solutions. A Pareto optimum is a point in Ω_t where it isn't possible to

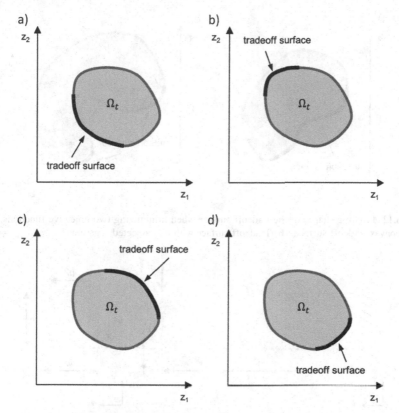

Fig. 6.11 Different types of Multi-Objective Optimization with two objective functions. (**a**) Minimize z_1, minimize z_2. (**b**) Minimize z_1, maximize z_2. (**c**) Maximize z_1, maximize z_2. (**d**) Maximize z_1, minimize z_2

improve one goal without decreasing another goal at the same time (Coello Coello 1999). The set of Pareto solutions after the Multi-Objective Optimization is called the tradeoff surface. In Fig. 6.11, we see an abstract set of solutions Ω_t plotted in the (z_1, z_2)-plane. The subplots show the tradeoff surface for the different kinds of optimization.

Tradeoff surfaces can assume many different shapes. The simplest one is the convex surface shown in Fig. 6.11. But it is also possible that the Pareto-surface is not convex (Fig. 6.12a) or consists of unconnected segments (Fig. 6.12b).

Example 6.2 As an example, we use a hollow beam under a given load F. The optimization problem is shown in Fig. 6.13. As optimization parameters, we use the height p_1 and the width p_2 of the rectangle inside of the hollow beam. The outside dimensions h, w and l are constant during the optimization.

The goal is to minimize the mass m, as well as to minimize the maximum displacement d of the hollow beam under the load F.

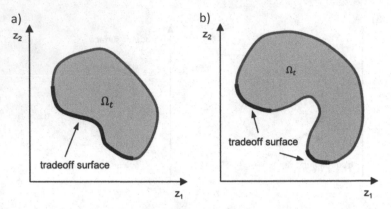

Fig. 6.12 Possible shapes of the tradeoff surface when minimizing two objective functions. (**a**) Non-convex tradeoff surface. (**b**) Tradeoff surface with unconnected segments

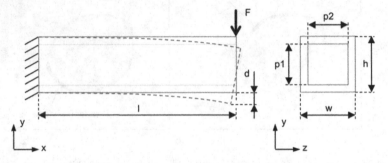

Fig. 6.13 The structure of the hollow-beam problem

The mass is calculated by

$$z_1(p_1, p_2) = m(p_1, p_2) = (w \cdot h - p_1 \cdot p_2) \cdot l \cdot \rho.$$

The maximum displacement d is calculated by

$$z_2(p_1, p_2) = d(p_1, p_2) = \frac{F \cdot l^3}{3 \cdot E \cdot I_y(p_1, p_2)},$$

$$\text{with } I_y(p_1, p_2) = \frac{w \cdot h^3 - p_2 \cdot p_1^{3}}{12}.$$

Our design space is restricted by the upper and lower limit of the two input parameters. The range is

$$10 \leq p_1, p_2 \leq 14.$$

The constant values for geometry and material are shown in Table 6.1.

Table 6.1 Input data used in the hollow-beam problem

Constant	Value
E	2.1×10^5 N/mm^2
ρ	7.85 g/cm^3
w	15 mm
h	15 mm
l	100 mm
F	100 N

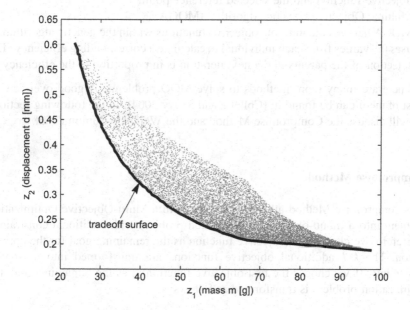

Fig. 6.14 Tradeoff surface of the hollow-beam problem

In Fig. 6.14 we see the set of possible results for the objective functions within the parameter range. To generate this set, we randomly choose some values for the parameters with $p_1, p_2 \epsilon [10, 14]$ and calculate the values of the objective functions. The set of solution is plotted in the (z_1, z_2)-plane. In this example we can see that there is not one singular solution with a minimum weight z_1 and a minimum displacement z_2.

6.2.2   Strategies for MOO

In the previous example, we found the tradeoff surface by calculating many designs and extracting the tradeoff surface from the results. This method is not very efficient because we get many designs which we are not interested in. To calculate the Pareto optimal points or the tradeoff surface directly, there are many different methods available. The following list shows the most used methods to calculate the tradeoff surface.

- Compromise Method
- Weighted-Sum
- Distance-to-a-reference-objective Method
  In this method we define a reference point with values for each objective function. The new goal is to minimize the distance between the result of the objective function and the selected reference point.
- Multiple Objective Genetic Algorithm (MOGA)
  MOGA handles the multiple objective functions within the genetic algorithm. It uses the values from each individual to calculate a corresponding efficiency. The selection of the parents in the next iteration is in proportion to the efficiency.

There are many more methods to solve MOO problems. A good overview of most of them can be found in (Collette and Siarry 2004). In the following sections we will discuss the Compromise Method and the Weighted Sum in detail.

Compromise Method

The Compromise Method allows us to transform a Multi-Objective Optimization problem into a mono-objective optimization problem with additional constraints. Therefore, we choose one objective function as the remaining goal for the optimization. The $k-1$ additional objective functions are transformed into inequality constraints. If we choose the first objective function z_1 as the remaining goal, the optimization problem is transformed as follows:

$$
\begin{aligned}
\text{minimize } & z_1(\mathbf{p}) \\
& z_2(\mathbf{p}) \leq \varepsilon_2 \\
& \vdots \\
& z_k(\mathbf{p}) \leq \varepsilon_k \\
& \mathbf{g}(\mathbf{p}) \leq 0
\end{aligned}
$$

For an optimization task with initially two objectives, this new formulation leads to an optimization problem visualized in Fig. 6.15. Here the second objective function z_2 is constrained by the value ε_2. The goal is to minimize the objective function z_1. As result we obtain one Pareto point $z_{1,min}$.

For the identification of other Pareto points and to obtain a tradeoff surface with this method, we perform multiple optimization runs and vary the value ε_2 of the restricted objective function z_2.

Example 6.3 To get an idea how the tradeoff surface is calculated in a real problem, we use the hollow-beam problem introduced in Example 6.2. The objective function z_1 is defined as the goal. The objective function z_2 is transformed into a constraint. As we expect displacement values d in the range of $0 \ldots 0.6$ mm within

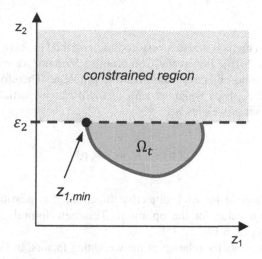

Fig. 6.15 Behavior of the compromise method

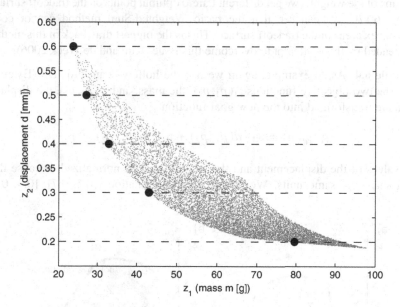

Fig. 6.16 Compromise method for the hollow-beam problem

the defined parameter range, we choose five values 0.2, 0.3, 0.4, 0.5 and 0.6 for the constraint value ε_2. In Fig. 6.16, we see the Pareto optima for the five constraints. We realize there is one severe disadvantage of this method. Due to the shape of the tradeoff surface, there is no point calculated between the mass $m = 40$ and $m = 80$, so we have no idea about the Pareto front in this region.

This might be resolved by switching the goal and restricting the objective z_1 (the mass of the hollow beam) with values ε_1 in the range of $m = 40 \ldots 80$.

Weighted Sum

The Weighted Sum method is also a very common method to solve MOO problems
(Marler and Arora 2010). Just as the Compromise Method, we try to convert the
problem into a mono-objective optimization problem. Therefore, we build a
resulting objective z_{eq} by a weighted sum of the different partial objectives and
by an appropriate set of weights w_i.

$$z_{eq}(\mathbf{p}) = \sum_{i=1}^{k} w_i \cdot z_i(\mathbf{p}) \tag{6.14}$$

By adjusting the weight for each objective function, it is possible to define the
importance of each value for the optimum. This new formulation leads to the
behavior shown in Fig. 6.17.

The line L_1 represents the relation of the weighting factors. In Fig. 6.17a, we get
one unique solution, here with an equal weight for both objective functions. By
variation of the weights, we get different Pareto optimal points on the tradeoff surface.
In Fig. 6.17b, we can see that the basic Weighted-Sum method cannot cover
non-convex areas of the tradeoff surface. This is the biggest drawback of this method
so extended methods attempt to overcome this issue (Kim and de Weck 2006).

Example 6.4 As an example, again we use the hollow-beam problem (Example
6.2). The two objective functions $z_1(p_1,p_2)$, the mass, and $z_2(p_1,p_2)$, the displacement, are transformed into the new goal function

$$z_{eq}(\mathbf{p}) = w_1 \cdot d(p_1, p_2) + w_2 \cdot m(p_1, p_2).$$

The values of the displacement and the weight must be normalized, because they
don't share the same units. We choose five combinations [0.3, 0.7], [0.4, 0.6],

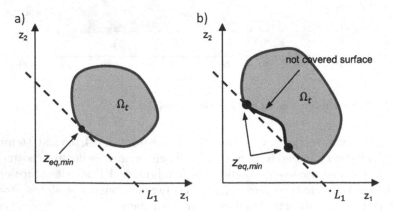

Fig. 6.17 Behavior of the Weighted-Sum method. (a) Convex tradeoff surface. (b) Non-convex
tradeoff surface

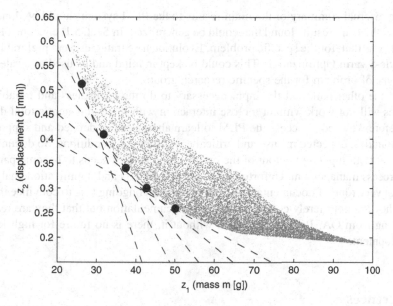

Fig. 6.18 Weighted Sum method for the hollow beam problem

[0.5, 0.5], [0.6, 0.4] and [0.7, 0.3] for the weights $[w_1, w_2]$. In Fig. 6.18 we see the resulting Pareto optima.

6.3 Optimization and Process Management of the Virtual Development Process

Rolf Steinbuch

Among the most important but most troublesome tasks in CAE is the management of large amounts of data. Increased Quality Assurance (QA) requires the documentation of every step, every component and every detail of the virtual product development and the real lifetime of a system. This affects the design process as well. Because the optimization is included in the design, the optimization process together with all its assumptions has to be documented as well. But it is impossible to collect all the ideas designers had while working during the virtual development. A collection of all misleading ideas would only add to the overflow of stored data, which nobody is ever going to look at again.

But all this searching and trying and pursuing misleading directions creates a rich experience-based knowledge that should be available to the next subsequent projects. Design teams are supposed to produce a history of what to do, when and why. We are not discussing if this needs to be integrated in the Product-Lifecycle-Management (PLM) systems or not. But not building up a system of knowledge leads teams to repeated errors that could be easily avoided.

An external summary of the main results to the PLM system should be done as soon as there are results found that could be generalized. In Sect. 5.1, e.g., one of the results was that, for the specific problem, Evolutionary Strategies were preferable to Particle-Swarm Optimization. This could be kept in mind and be used as a rule for this type of problem for the specific research group.

On the other hand, all the input necessary to do the robustness and reliability studies will not work without a close interaction with the component's total data. Therefore we need to access the PLM to learn about scatter, defined and supposed uncertainties, expected misuse and critical environmental conditions. So documentation of both input and output of the optimization studies needs to become part of the process management. Unfortunately, many designers and optimization analysts are not very fond of documentation. So it remains an ongoing task to convince them that they are not merely contributing to the documentation but that they are really profiting from QA. It is worth it, and without it, there is no future for high level development.

References

Coello Coello, C. A. (1999). An updated survey of evolutionary multiobjective optimization techniques: state of the art and future trends. In *Proceedings of the 1999 Congress on Evolutionary Computation, 1999, CEC 99* (Vol. 1, pp. 13). doi:10.1109/CEC.1999.781901.

Collette, Y., & Siarry, P. (2004). *Multiobjective optimization: Principles and case studies*. Berlin: Springer.

Du, X., & Chen, W. (2004). Sequential optimization and reliability assessment for probabilistic design. *ASME Journal of Mechanical Design, 126*, 225–233.

Du, X., Sudjianto, A., & Chen, W. (2004). An integrated framework for optimization under uncertainty using inverse reliability strategy. *ASME Journal of Mechanical Design, 126*, 562–570.

Gekeler, S., & Steinbuch, R. (2014). *Remarks on robust and reliable design optimization*. METAHEURISTICS AND ENGINEERING, 15th Workshop of the EURO Working Group. S. 77–82.

Kim, I. Y., & de Weck, O. L. (2006). Adaptive weighted sum method for multiobjective optimization: A new method for Pareto front generation. *Structural and Multidisciplinary Optimization, 31*, 105–116.

Marler, R. T., & Arora, J. S. (2010). The weighted sum method for multiobjective optimization: New insights. *Structural and Multidisciplinary Optimization, 41*(6), 853–862.

Tu, J., Choi, K. K., & Young, H. P. (1999). A new study on reliability-based design optimization. *ASME Journal of Mechanical Design, 121*, 557–564.

Wang, Z., Huang, H., Liu, Y. (2010). A unified framework for integrated optimization under uncertainty. *ASME Journal of Mechanical Design, 132*(5), 051008-1–051008-8.

Yin, X., & Chen, W. (2006). Enhanced sequential optimization and reliability assessment method for probabilistic optimization with varying design variance. *Structure and Infrastructure Engineering, 2*, 261–275.

Chapter 7
Future Tasks in Optimization and Simulation

Simon Gekeler and Rolf Steinbuch

Broad acceptance of Finite-Element-based analysis of structural problems and the increased availability of CAD-systems for structural tasks, which help to generate meshes of non-trivial geometries, have been setting a standard for the evaluation of designs in mechanical engineering in the last few decades. The development of automated or semi-automated optimizers, integrated into the Computer-Aided Engineering (CAE)-packages or working as outer loop machines, requiring the solver to do the analysis of the specific designs, has been accepted by most advanced users of the simulation community as well. The availability and inexpensive processing power of computers is increasing without any limitations foreseen in the coming years. There is little doubt that Virtual Product Development will continue using the tools that have proved to be so successful and so easy to handle.

7.1 Main Trends in Optimization

Focusing on optimization alone, we could conclude that the doors opened in the last few years have enabled many fruitful developments. We feel that the combination of stochastic methods and optimization, such as reliability and robustness, will play a major role in science and engineering in the next years. Not only can these methods analyze broader ranges of a part's design, but they can also estimate loading over a lifetime as well. In consequence, the real loading situation will be better understood. This understanding includes the response of aged parts—parts that have been subject to long-term mechanical or corrosive loading, including wear resulting from long continuous loads and the degeneration of the materials.

S. Gekeler • R. Steinbuch (✉)
Hochschule Reutlingen, Reutlingen Research Institute, Alteburgstraße 150, 72762 Reutlingen, Germany
e-mail: Simon.Gekeler@Reutlingen-University.DE; Rolf.Steinbuch@Reutlingen-University.DE

© Springer-Verlag Berlin Heidelberg 2016   147
R. Steinbuch, S. Gekeler (eds.), *Bionic Optimization in Structural Design*,
DOI 10.1007/978-3-662-46596-7_7

Even the amount of time and the power-consumption problem of Multi-Objective Optimization (MOO, cf. Sect. 6.2) will yield reliable and fast procedures. Whether it will be weighting of different goals or switching between goals and restrictions, any further developed method must be evaluated by its efficiencies in the industrial and scientific applications. On the other hand, there might yet be an increased understanding of the MOO character of many engineering decisions, so that even better decision-making criteria could be used.

One expected, important development in optimization is the deepening of the applications. Today we are mostly concerned with the shape of a given part or assembly. The integration of the solution for the design of the part and the design methodology do not yet include the potential of Virtual Product Development. However, the manufacturing methods available and the life-time history, including the setup, are not yet within the scope of most optimization studies today. While an expansion of the optimization process might be observed in this field, this might develop in parallel to the stochastic approaches mentioned above.

Furthermore, optimization will be part of nearly all design processes. This even includes studies of low-cost parts such as cans or bottles, which are not dealt with today. Everyone uses them, but no one knows about the number of manufacturing steps and corresponding optimization parameters.

A simple everyday example is an aluminum can. Currently to manufacture the bottom, about four manufacturing processes with more than 10 free parameters steps are required (Fig. 7.1a). But to produce the particular shape of the top of the

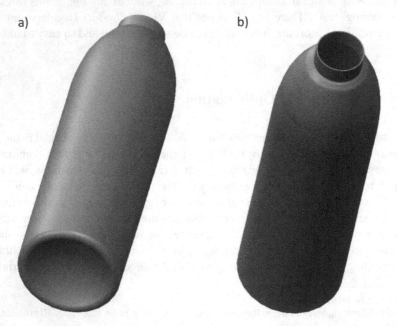

a)   b)

Fig. 7.1 Simple can as a high-end optimization problem. (**a**) Bottom defined by 10 parameters. (**b**) Upper section defined by more than 200 parameters

can, we need a repeated reduction of the radius in about 20 steps (Fig. 7.1b). Every forming tool of these 20 steps is defined by at least 10, if not more, parameters. Just to model the manufacturing of this simple can, we need to optimize a problem with about 400 free parameters if we include the additional steps of the process. Looking back at Fig. 3.2, where we plotted the number of variants required to do an optimization as a function of the free parameters, we come to $ntrial > 10^6$ variants. As we realize that each job or study of one can design takes more than 1 h, we realize that our current approaches will not be sufficient to solve such problems, even when using massive parallel computing. New methods of dealing with optimization are required. But there should be no doubt that they will be found.

But—not disregarding this need for more powerful optimizations covering larger aspects of the product's history—optimization will continue to be available to more and more users. Integration of all CAE-processes will allow even inexperienced engineers to do highly skilled studies. Optimization will be integrated into the daily designer's work.

7.2 Qualifications and Quality Assurance

The expansion of the community using advanced CAE-systems will have an impact on optimization as well. First, more and less qualified engineers will be using these systems, drawing their own conclusions and deriving real designs from them.

The inherent danger in this growth is the misinterpretation of the results: e.g. the importance of a sufficient high reliability or robustness of the design may cause an inexperienced engineer to have a too optimistic estimation of his or her proposal. We all know that we do not read all the warnings a computer code or a Graphical User Interface (GUI) gives after a solution has been found. Why should designers with little background in simulation and optimization read all the comments which they don't really understand and that may only confuse them?

The automation of CAE-processes therefore needs a highly evolved Quality Assurance (QA) strategy that prevents misunderstanding of the proposed solutions found by processes which run without much understanding and much input from the users. But how do we design a warning system to inform inexperienced users that some appealing proposals are not recommended and to guarantee that the users understand the importance of that statement?

Different institutions like NAFEMS, with a Professional Simulation Engineer certification   (http://www.nafems.org/professional_development/certification/, retrieved 15.04.2015) or universities offer qualifications not only to students, but to professionals as well. They focus not only on teaching more and more methods to solve problems but also careful and critical reviews of the results. Perhaps it would be a good idea to require some kind of formal qualification for all people who are responsible for important decisions related to simulation results. The experience of the authors and many other people involved in simulation proves however, that there is not much support in industry for such proposals.

7.3   Interpretation of Simulation Results

Another problem not only related to optimization is the implicit potential of CAE-systems to provide results without qualified input. To illustrate this danger, we give a simple example. There are CAE-systems that do stress analysis studies without requiring much understanding of the users. When asked for lifetime predictions, these systems use the tensile stress and the surface properties to predict the fatigue properties of the material and produce approximations of the component's expected lifetime. Everyone who has ever worked with fatigue problems knows that there are more contributing factors related to lifetime prediction than just ultimate stress and surface quality. Nevertheless, some people use such user friendly systems and rely on their results, without considering their validity. Consequently, it is no surprise that the predicted lifetimes are not very close to the real ones.

To avoid such nonsense-results, there should be a requirement for the software supply companies to qualify the methods they offer. But who has the qualifications and resources to decide which approach is reliable? We once did a survey of different commercial systems to predict the lifetime of parts using available simulation results. Experienced engineers would not be surprised to hear that the predicted lifetimes differed by a factor of 10,000. This corresponds to the difference between 1 h and 1 year.

Here, as in other engineering applications, the standardization of simulation methods seems to be inevitable. But who will develop the standard? We feel that the broader the simulation application becomes, the more conflicts will spring up over the creation of such standards.

7.4   Believing in Standards and Defaults

We can easily expand the dangers resulting from inexperienced users of simulation tools to users of optimization systems. How do we convince engineers working in shops to repeat their studies, to use bionic methods, to do many variants, and many generations when available optimizers integrated into CAE-systems produce promising results within short times? Why should they do reliability and robustness analysis, when they don't know much about the scatter of the properties of the parts to be assembled? We are often confronted with either ignorance or misunderstanding if we ask for more qualified research.

Here the approximations available through the optimization process might be used to provide some idea if the solution found really is close to the best one, or if there are reasons that other regions of the parameter space should be examined as well. Structures of meta-modeling might be used to estimate the responses in the unexplored range.

Nevertheless, the curse of the dimensions (cf. Sect. 3.2) and the always limited time to produce proposals tend to prevent such additional searches. Perhaps automated systems that are more sensitive in less efficient regions of the solution space will be able to propose and to perform such complementary processes during times when there is free computing power. But experience up to now leads to the expectation that the tendency of users to accept everything proposed by the "system" will remain. Even worse, users are crying for yet more defaults, while fewer questions are expected to be asked of them.

7.5   Linking Development and Manufacturing

Another important step that is expected in the evolution of optimization is the connection between development and manufacturing. Virtual plants help to create production lines that are close to optimal. But in many studies, the requirements of the manufacturing process are not examined during the design process. The integration of design and production optimization could be an essential step towards CAE-integrated development. In Sect. 7.1 we already mentioned such a problem. The deepening use of CAE-tools should help to provide a better integration of the total product genesis, from the idea to its creation in the final manufacturing process.

The word of design according to the material is well known. We should expand it to an integrated virtual development from the function of the part or system to the manufacturing and the materials selection.

7.6   New and Old Materials

The use of new materials also needs to be integrated into the design stage. We use material data bases for metals and many non-metallic materials. But for composites, there is much more data required than for more or less isotropic and homogenous materials. Here, some software developers, such as Digimat (http://www. mscsoftware.com/de/product/digimat, retrieved 15.04.2015), are attempting to establish methods to learn about the material response by modelling the manufacturing process of such composites and then exporting these local material models to the simulation codes.

On the other hand, we should be aware that many elements of material response are not yet fully understood. If we check what material departments at our universities are researching, we find that many of them are working with metal fatigue problems. This implies that we, the human race, have been using metals for about 3000 years, but do not understand why and how metals fail after some time of service. How then could we be expected to understand the response of much more complicated materials such as composites, fibers, or plastics?

Concluding that we are not able to do any reliable prediction of a material's load carrying capacity is not the point. But we need to be aware of the difficulties included in the interpretation of any stress analysis result for any material. Blind faith in answers provided by automated systems is not the safe way. The interpretation of material responses is still a task that must be handled with care and responsibility, especially in the case of optimization, when hundreds or thousands of variants are checked in a loop which no human being will ever fully examine.

7.7 Reliable Loading Systems

The extent and accuracy of loading systems, which is always a difficult point in the setup and interpretation of numerical structural mechanics, might improve as the importance of stochastic processes becomes more and more accepted by the community. Whether this will yield more reliable and less biased views of the loading is not yet clear. Nevertheless, all training on, and all discussion of, simulation results should take into account that only realistic loads will help find the system's real response.

The availability of databases containing real loading systems, the possibility to link such databases to stochastic processes, and increasing computing power will help to find proposals that better describe the real loads acting on a part or a system. But here nothing helps except going out in the field and learning what real current customers do with our products. The simulation and the optimization need to reflect not how we think customers use the products, but rather how they really use them. Especially the changes of users' habits may completely alter the landscape. For example, change in the typical ownership of relatively expensive cars may underline the importance of realistic reviewing of loading. Twenty to thirty years ago, expensive cars were mostly driven by older customers, who took painstaking care of their precious vehicles. Today, an increasing number of these expensive cars are owned by younger consumers with a great deal of money, but who don't realize how critical is the regular maintenance of their vehicle. Therefore, the loading system has had to be re-designed in cases where the most typical owner is much younger than previously was the case.

7.8 Preprocessing and Meshing

New meshers and pre-processor generations have been introduced in the past few years. Whether they really produce better meshes and so enable simulation analyses on available computers in acceptable time is not yet sure. Currently, automated meshers often produce fine meshes in regions where there is no need for this precision, while creating meshes at some critical regions where the elements are far too coarse. The adaptive meshing available in many codes may help, but it may deceive users that numerical results are reliable, because the automated re-meshing guarantees numerical convergence.

Future systems need to find acceptable meshes and corresponding element types with a high reliability. For optimization jobs, this is of special interest. If small differences in the results of a simulation are caused by numerics and not by the variation of the design, misleading search directions could be stimulated. This is especially true in regions where the gradient of the goal and the constraints is very flat, while the sensitivity to numerical disturbances is rather high. New automated systems should be able to provide counter measures to avoid such numerical traps.

All these simulation problems influence the power of optimization tools even more. If you have to reconfigure your material definition for each variant, such as in the case of composites, the computing time per individual or variation will substantially increase. The total conceptualization of the simulation environment has to be expanded by additional pre- and post-processing codes.

The acceleration necessary to perform optimization in a reasonable time is overwhelmed by the larger scope of individual jobs. On the other hand, more and more computing power is becoming available, so parallelization, in all its different appearances, will help to reduce the calculation time for the many individual jobs of Bionic Optimization studies.

Even if the difficulties mentioned do not vanish within a short time, and without sufficient discussion about the necessity of qualified application of high-level tools, there should be no doubt that the methods of Bionics and the upcoming trend of stochastic interpretation of real processes will lead to a new understanding of the total design of parts and assemblies. The fact that optimal designs often have a certain similarity to structures we find in the nature is one of the most inspiring insights to designers. Why don't we send our designers into a forest to find promising initial ideas? Why don't we include the ability of natural designs to withstand very different loading conditions, from strong winters while the ground is soaked to dry summers without enough humidity, to allow a required ductility into our conceptualization of designs? There is an old German engineering proverb that states, "A good design is well shaped, adapted to materials properties, and lightweight." Does this proverb not include many of the ideas of reliable and robust Bionic Optimization? Even if there are still many things to be learned, there should be no doubt that Bionics are one of the most promising directions of the future development of engineering.

Index

© Springer-Verlag Berlin Heidelberg 2016
R. Steinbuch, S. Gekeler (eds.), *Bionic Optimization in Structural Design*,
DOI 10.1007/978-3-662-46596-7

Printed in the United States
By Bookmasters